Fundamentals *of* Air Sampling

Fundamentals of Air Sampling

Gregory D. Wight

Professor of Environmental Technology
Civil Engineering Department
Norwich University
Northfield, Vermont

CRC Press
Taylor & Francis Group
Boca Raton London New York

CRC Press is an imprint of the
Taylor & Francis Group, an **informa** business

CRC Press
Taylor & Francis Group
6000 Broken Sound Parkway NW, Suite 300
Boca Raton, FL 33487-2742

First issued in paperback 2020

ISBN 13: 978-0-367-57979-1 (pbk)
ISBN 13: 978-0-87371-826-4 (hbk)

Visit the Taylor & Francis Web site at
http://www.taylorandfrancis.com

and the CRC Press Web site at
http://www.crcpress.com

This book is dedicated to my students—
your success is my reward.

PREFACE

The field of air pollution control is a rapidly changing one. The changes, however, have often meant discovery of new challenges, rather than final solution of long known problems. This is certainly true in the area of air sampling, where long-ago developed techniques must now be adapted to newly recognized problems of volatile compound emissions and air toxics sampling. The Clean Air Act Amendments of 1990 put even more emphasis on traditional emission sampling, with significant new Title IV Acid Deposition Control requirements and Title V Operating Permit requirements.

There will be a need for trained air quality measurement personnel in government and industry for many years to come. It is the aim of this text to provide some fundamental background in this field. The book is written for engineers and technologists who have had some background in fluid mechanics and energy issues. College students can follow the text, as my students have, through a one semester course. Practicing engineers and air sampling personnel will find the book a useful reference.

Both SI and US Customary units are used in this text. In spite of the federal government's efforts to stress SI units first, US Customary units are still in widespread use, particularly in the combustion field, and air sampling personnel will need to be conversant in both. Extensive units conversion tables are included in Appendix A.

I would like to acknowledge the input of my students at Norwich University into the development of this text. Students from many years, many now practicing in the air measurement industry, have enthusiastically prodded, questioned, and provided unique insights into problem solutions and text development.

Many people at Norwich University and its Engineering Division have been most supportive in the development of this book. In particular, thanks are due to Jackie Hatch for her cheerful and capable clerical support and to colleagues Eugene Sevi, Alan Fillip, and Jon Allen.

Many of the photographs were taken by my son Michael. Above all, I want to acknowledge my wife Tammy, a steadfast supporter throughout the long extra hours required to develop this book.

THE AUTHOR

Gregory Wight is Professor of Civil and Environmental Engineering at Norwich University in Northfield, Vermont. After an undergraduate degree in Mechanical Engineering from MIT, his career in Air Quality began with graduate work and a PHS Traineeship at the University of Florida.

The Clean Air Act Amendments of 1970 and EPA state grants brought him to the Connecticut Department of Environmental Protection in the early 1970s, when everyone was positive air pollution would be conquered by 1975. An opportunity to teach in the area of air quality and emission measurement at Norwich was the long-ago genesis of this text. He has also developed undergraduate air quality courses for the Environmental Engineering Department at the United States Air Force Academy, and served as a Distinguished Visiting Professor at the Academy.

A long-time member of both the American Society of Mechanical Engineers and the Air and Waste Management Association (AWMA), he has served on the Board of Directors of the New England Section of the Air and Waste Management Association, and founded the Norwich student chapter of AWMA.

CONTENTS

Fundamentals of Air Sampling

1 INTRODUCTION TO AIR SAMPLING

Since the earliest versions of the U.S. Clean Air Act in the 1960s established maximum safe pollutant concentrations, scientists and engineers have been developing methods to quantify air pollutant concentrations and rates of release of pollutants into the air from sources. Early crude methods such as smoke density assessment tools (Ringelmann Charts), dustfall buckets, and soiling index have largely been replaced by continuous emission monitors, inhalable particulate samplers, Fourier Transform Infrared detectors (FTIDs), cryogenic trapping sampling and analysis techniques, etc. The essentials, however, of pollutant collection technologies have not changed. To quantify air pollutant levels, the pollutant concentration must be accurately known. In many air quality or emission assessment applications, concentration is found as the quotient of the mass of the pollutant of interest and the volume of air or gas sampled that contained that amount of pollutant.

The quantification of pollutant mass is primarily the job of the analytical laboratory or of the lab personnel. The quantification of the volume of gas sampled, however, is the job of the air sampling or field personnel. It is this air sampling portion of the overall air quality assessment process that this text addresses.

There are hundreds of techniques and types of instruments that allow the quantification of mass of collected air pollutant — some that require a few milligrams for accurate assessment, like Total Suspended Particulate samplers, others that respond to fantastically small quantities of the pollutant (e.g., gas chromatography/mass spectroscopy, (GC/MS), perhaps even in the range of a few molecules (e.g., alpha track radon detectors).

The volume of air or exhaust gas in which the pollutant of interest is mixed must be as accurately measured as the pollutant mass. A number of volume, gas velocity, and flow rate measuring or controlling devices are employed in the process of air and gas sampling. The choice of technique depends greatly upon the properties of the gas to be sampled and the pollutant to be collected. Fundamental concepts of physics, thermodynamics, and fluid mechanics form the foundation for calibration and use of these devices and for computation of gas volume.

This text concentrates primarily on methods to accurately collect the air or gas sample and accurately quantify the volume of gas collected.

The term *air sampling* as used here refers to five distinct sampling categories:

1. Source or emission sampling — the measurement of gas flow rate, physical characteristics, and composition and pollutant concentration in exhaust gas streams leaving a process, factory, chimney, or ventilation system and entering the atmosphere.
2. Ambient sampling — the measurement of outdoor air pollutant levels, generally in attempts to characterize fairly broad area pollutant levels to assess health effects or to forecast effects of a proposed new source of pollutants.
3. Industrial hygiene air sampling — usually indoor measurement in workplaces, either stationary or area monitoring or personal monitoring (carried by an employee) for worker protection.
4. Residential indoor sampling — to evaluate healthfulness of indoor air in living areas.
5. Process or quality control — gas sampling in order to monitor the makeup of a production process or a manufactured product.

Standards — maximum allowable concentrations of contaminants in the air or gas — are set by the Environmental Protection Agency (EPA), state and local environmental agencies, manufacturers, or professional organizations. The American Society for Testing and Materials (ASTM), American Society of Mechanical Engineers (ASME), American Council of Governmental Industrial Hygienists (ACGIH), American Society of Heating, Refrigeration, and Air Conditioning Engineers (ASHRAE), and governmental agencies like the Occupational Safety and Health Administration (OSHA) or its counterparts at the state level, and the National Institute of Occupational Safety and Health (NIOSH) as well as numerous private consultants have participated in the development of standards and refinement of methods for sampling air and gas over the past 30 or more years.

An important distinction should be made between air sampling and air monitoring. This book stresses techniques for *sampling* — the proper techniques to acquire a representative sample of possibly contaminated air or gas for later evaluation or analysis of the quantity of pollutant present. *Monitoring*, on the other hand, generally refers to methods and equipment designed to give real-time or near-immediate information on air or gas characteristics, especially pollutant concentrations. Continuous or intermittent instruments are in wide-ranging use as stack gas monitors (continuous emission monitors, CEMS[48]) and as ambient, industrial, indoor, or process control monitoring devices. While the sampling concepts and techniques introduced in this book are essential for air and gas monitoring, considerable additional information, beyond the scope of this text, is needed for a full understanding of continuous air monitoring.

Air sampling, the collection of a sample for later analysis, has become a good deal more important in recent years due to three developments:

1. Widespread awareness of the dangers of even small quantities of organic or other toxic pollutants in the air
2. The development of toxic air pollutant regulations by state air pollution control agencies, under the guidance of the EPA
3. The development of highly sophisticated laboratory techniques for measurement of minute quantities of pollutants collected from air or exhaust gas streams

As a result, there is more demand for air sampling, particularly emission sampling for toxic pollutants and indoor sampling. As well, there is more public scrutiny, and a consequently greater need for accuracy and carefully documented procedures and quality assurance programs.[19]

Continuous monitors often do not have the required sensitivity to measure toxic air pollutants at very low concentrations. Sampling techniques must be used to collect and, in some cases, concentrate a sample for later laboratory analysis by gravimetric techniques or by GC/MS, flame ionization detection (FID), photoionization detection (PID), electron capture detection (ECD), or atomic absorption spectrometry (AA), ion chromatography, or other techniques.[22,30]

The collection of a representative sample of contaminated air or gas requires attention to a number of parameters of the gas: for example, temperature, pressure, major constituent volume percentages, average molecular weight, humidity, and sampling rate, as well as volume collected. Chapter 2 provides a brief review of some important concepts and their application to air sampling.

2 BASIC GAS CONCEPTS

2.1 INTRODUCTION

This chapter provides a brief review of some essential principles of physics, thermodynamics, and fluid mechanics.

2.2 PRESSURE MEASUREMENT

Fluid pressure measurement is nearly always made relative to the surroundings or atmospheric pressure by a mercury or water manometer, or by a pressure transducer or mechanical gauge. Such relative pressure is called "gage" pressure (while "gauge" is the preferred spelling, "gage" is an accepted alternate) and is positive if higher than surroundings, negative if lower. Relative to atmospheric pressure, a negative gage pressure is called "vacuum".

The most commonly used units for gage pressure are pounds per square inch, gage (psig), inches or millimeters of mercury or water, or Pascals or kiloPascals (Pa, kPa; 1 Pa = 1 Newton per square meter.) See Table 2.1 for conversions.

Table 2.1 Pressure Units

1 atm = 14.69 psi (lbf/in^2)	
= 760 mm Hg	
= 101.325 kPa	
= 1013.25 mB	
= 29.92" Hg	density ratio: $\rho_{Hg} = 13.6\ \rho_{H_2O}$
1" Hg = 13.6" H_2O	
= 25.4 mm Hg	
= (13.6 × 25.4) = 345.44 mm H_2O	

Gage pressure is used because it is easy to measure, record, and report; but for any computation involving fluid pressures, the actual or "absolute" pressure is needed, rather than the relative pressure. Since gage pressure is that amount above or below the surroundings, it is obvious that absolute pressure is the sum of surrounding and gage pressures.

The surrounding pressure, in the case of air sampling measurements, is nearly always the local atmospheric or barometric (that which the barometer reports) pressure. Mercury or aneroid barometers are used to measure atmospheric pressures and commonly used units are pounds per square inch, absolute (psia), kPa, inches of mercury (" Hg), millimeters of mercury (mm Hg), or millibars (mB).

$$P_{abs} = P_{bar} + P_{gage}$$

$$\text{or} \quad P_{atm} + P_{gage} \quad \text{in consistent units} \tag{2.1}$$

Gage pressure, recall, may be positive or negative; thus, P_{abs} may be greater or less than P_{atm}. Thus, a gauge that reads 50 mm Hg vacuum when barometric pressure is 750 mm Hg yields an absolute pressure of 700 mm Hg.

Gage pressure is also sometimes called "pressure head" or just "head". A gas tank with a head of 4 psig means 4 psig higher pressure inside the tank relative to the air outside the tank. A flow restriction in a gas or fluid stream is measured in terms of "head loss" or pressure change from upstream to downstream. Head can be measured in terms of any fluid — usually water or mercury — and is related to other pressure units by:

$$P = \rho \times h \times (g/g_c)$$

where P = pressure, in lbf/in^2 or other appropriate units
ρ = density of liquid, e.g., lbm/in^3
h = head, e.g., inches
g = gravitational acceleration in ft/sec^2
g_c = constant = 32.2 $lbm\text{-}ft/lbf\text{-}sec^2$

2.3 TEMPERATURE

Fluid or gas temperature is measured with thermometers or thermocouples. The measured values will be in Fahrenheit (°F) or Celsius (°C). For calculations, absolute temperatures must be used, in Rankine (°R) or Kelvin (K). See Table 2.2.

Table 2.2 Temperature Units

°R = °F + 459.67
°K = °C + 273.15
°R = K × 1.8
°C = (°F − 32)/1.8

Example Problem 2.1

The pressure gauge on a helium tank reads 200 psig. A mercury barometer in the same room indicates the current barometric pressure. What is the tank absolute pressure in psia if the barometric pressure is:

a) 768 mm Hg?
b) 29.7" Hg?
c) 1030 mB?

Solution

a) 768 mm Hg × 14.69 psi/760 mm Hg + 200 = 214.8 psia
b) 29.7" Hg × 14.69 psi/29.92" Hg + 200 = 214.6 psia
c) 1030 mB × 14.69 psi/1013.25 mB + 200 = 214.9 psia

Example Problem 2.2

A vacuum pump draws air through an air sampling device. Just upstream of the pump, a vacuum gauge is connected and reads 32" water vacuum. The barometer reads 751 mm Hg. What is the absolute pressure of the air at this location in:

a) mm Hg?
b) kPa?

Solution

a) (−32" H_2O × 1" Hg/13.6" H_2O) × 25.4 mm/in + 751 = 691.2 mm Hg
b) From a), absolute pressure = 691.2 mm Hg
691.2 × 101.325 kPa/760 mm Hg = 92.2 kPa

Example Problem 2.3

A column of water 10" high is supported by the pressure in a tank of gas. What is the gage pressure in psig?

Solution

The weight of the water is equal to the volume times the density times the acceleration of gravity, g; with the proper units:

$$\text{Weight} = \rho \times \text{vol} \times g/g_c = \rho \times h \times A \times g/g_c$$

where $h = 10"$

A = cross-sectional area of the water column

Pressure = $P = \text{Weight}/A = \rho \times h \times g/g_c$

For $g = 32.2\ \text{ft/sec}^2$ and $\rho = 62.4\ \text{lbm/ft}^3$,

$g_c = 32.2\ (\text{lbm-ft/sec}^2)/\text{lbf}$

$$P = \frac{62.4\ \text{lbm}/\text{ft}^3}{1728\ \text{in}^3/\text{ft}^3} \times 10 \times \frac{32.2}{32.2} = 0.361\ \text{psig}$$

2.4 IDEAL GASES

Ideal gases are those whose molecules do not attract one another and in which molecular collisions are elastic. All real gases deviate from these ideal behaviors to some extent. However, at the pressures and temperatures encountered in ambient air and exhaust gas measurement, the deviations are very slight, and little error is introduced in assuming ideal gas behavior.

The Ideal Gas Equation of State

The ideal gas equation of state, often called the Ideal Gas Law, is:

$$PV = mRT \tag{2.2}$$

where P = absolute pressure

V = volume of gas

T = absolute temperature

R = specific gas constant (as distinguished from the Universal Gas Constant, R_u; $R = R_u$/molecular weight)

m = mass of gas

A mole of a substance is the substance's molecular weight expressed in some mass unit (grams, kilograms, pounds mass) where the substance's molecular weight is the sum of the atomic weights of the atoms that compose the substance.

We know that:

$$\text{number of moles, n} = \frac{\text{mass}}{\text{molecular weight}} = \frac{m}{M} \qquad (2.3)$$

and

$$R_u = M \times R \qquad (2.4)$$

where R_u = universal gas constant
Therefore,

$$PV = mRT = \left(\frac{m}{M}\right) \times R_u T = nR_u T \qquad (2.5)$$

or

$$P\upsilon = RT \text{ or } \frac{P}{\rho} = RT \qquad (2.6)$$

where υ = specific volume = V/m
ρ = density = $1/\upsilon$

The units of R and R_u depend on the units used in the equation. Some useful values for R_u are listed in Table 2.3.

Table 2.3 Universal Gas Constant, R_u

0.08206 l-atm/gmole-K
62.4 l-mm Hg/gmole-K
8.314 m^3-kPa/kgmole-K = kJ/kgmole-K
1545.3 ft-lbf/lbmol-°R
0.7302 atm-ft^3/lbmol-°R
21.85" Hg-ft^3/lbmol-°R
10.73 psia-ft^3/lbmol-°R

Molar Volume

One kgmole of any ideal gas at 0 °C (273 K) and one standard atmosphere pressure (101.325 kPa, 760 mm Hg) will occupy 22.414 m^3. This constant is obtained from the Ideal Gas Law.
If n = 1 kgmole, P = 1 atm (101.325 kPa), R_u = 8.3143 KJ/(kgmole-K), and T = 273 K, then volume = (molar volume) = V_m

$$PV_m = R_u T \qquad (2.7)$$

or

$$(101.325 \text{ kPa}) \times V_m = (1 \text{ kgmole}) \times (8.3143 \text{ KJ/kgmole-K}) \times (273 \text{ K})$$

and

$$V_m = 22.414 \text{ m}^3 = \text{molar volume.}$$

To justify units in the above calculation, recall that kPa = KN/m^2 and KJ = KN × m, where KN = kiloNewtons, kPa = kiloPascals, KJ = kiloJoules, and m = meters.

Therefore, 1 kgmole of an ideal gas at 273 K and 760 mm Hg occupies 22.414 m^3, or 1 gmole occupies 22.414 L. Repeating, the molar volume of an ideal gas at 0 °C and 1 atm is:

$$V_m = 22.414 \text{ m}^3/\text{kgmole} \quad (2.8)$$

We know, however, that gases expand as temperature increases (Charles' Law, covered in the next section, gives the relationship); thus, at standard ambient conditions (760 mm Hg, 25 °C), 1 kgmole of any ideal gas will occupy a larger volume of 24.45 m^3 (or 1 gmole occupies 24.45 L).

$$V_m = 24.45\text{m}^3/\text{kgmole at } 25\ °\text{C, 1 atm} \quad (2.9)$$

This molar volume at 25 °C and 760 mm Hg pressure is of great importance in air sampling and air pollutant concentration calculations. From the Ideal Gas Law, two special cases also of great importance to air sampling work can be derived.

Boyle's Law

Named after Robert Boyle, Boyle's Law states that when the temperature of an ideal gas is held constant, the volume of a given mass varies inversely as the absolute pressure.

$$\frac{V_2}{V_1} = \frac{P_1}{P_2} \quad (2.10)$$

Charles' Law

Named after Jacques Charles, Charles' Law states that when the pressure of an ideal gas is held constant, the volume varies directly as the absolute temperature.

$$\frac{V_2}{V_1} = \frac{T_2}{T_1} \tag{2.11}$$

(Temperatures must be in absolute: Kelvin or Rankine.)

For example, if 5 L of an ideal gas at 1 atm and 20 °C are heated at constant pressure until T = 40 °C, the gas will expand to a volume V_2:

$$V_2 = 5 \times \frac{(40 + 273)}{(20 + 273)} = 5.34 \text{ liters}$$

Both Boyle's and Charles' Laws are often employed together in air sampling to find the equivalent volume of an ideal gas at a temperature and pressure different from actual.

$$V_2 = V_1 \times \left(\frac{P_1}{P_2}\right) \times \left(\frac{T_2}{T_1}\right) \tag{2.12}$$

Standard Temperature and Pressure

To be able to compare gas sampling data collected by various organizations or under differing conditions, all gas volumes must be corrected to a set of predetermined or *standard* conditions.

As noted earlier, for atmospheric or ambient sampling, the EPA has set these conditions as:

$$25\ °C = (298\ K) \text{ and } 1 \text{ atmosphere } (= 760 \text{ mm Hg})$$

For emission monitoring, EPA specifies Standard Temperature as 20 °C or 293 K. For example, if 3 L of ambient air are sampled at 35 °C and a pressure of 742 mm Hg, the sampled volume can be corrected to standard conditions by application of Boyle's and Charles' Laws (Equation 2.12).

$$V_{std} = V_{sampled} \times (P_{samp}/P_{std}) \times (T_{std}/T_{samp})$$

$$= 3\ L \times (742/760) \times [298/(35 + 273)] = 2.83 \text{ std liters}$$

2.5 GAS DENSITY

Another adaptation of the Ideal Gas Equation of State is for the calculation of gas density (density = mass/volume):

$$\text{density} = \rho = m/V = PM/R_uT$$

$$\text{or} \quad P\upsilon = RT$$

$$\text{or} \quad P/\rho = RT \tag{2.13}$$

where υ = specific volume = V/m, ρ = density, $R_u/M = R$

We can also find gas density from the standard molar volume of 24.45 m^3. If the gas pressure (P) and temperature (T) are measured in mm Hg and K, respectively, then

$$\text{density, } \rho = \frac{M \times \left(\frac{P}{760}\right) \times \left(\frac{298}{T}\right)}{24.45} \tag{2.14}$$

2.6 GASEOUS POLLUTANT CONCENTRATION

The concentration of any pollutant in air or exhaust gas is generally reported in units of mass per volume. For gaseous pollutants, units of volume of pollutant per volume of gas mixture are also used, and conversions from mass per volume to volume per volume employing gas law concepts are often needed.

From the Ideal Gas Law, the volume occupied by any ideal gas is:

$$V = m \times R_u \times \frac{T}{M \times P} \tag{2.15}$$

where m = mass
M = molecular weight

At standard atmospheric conditions, P = 760 mm Hg, T = 25 °C and, for V in milliliters (mL) and m in micrograms (µg), the pollutant volume is:

$$V\,(\text{mL}) = \frac{[m(\mu g)] \times 24.45 \times 10^{-3}}{M} \tag{2.16}$$

where the constant 24.45 will be recognized as the molar volume for any ideal gas at these standard conditions (Equation 2.9).

If the unit of gas volume into which the pollutant is mixed is 1 cubic meter (1 m^3), then both sides of the above equation can be divided by 1 m^3 to give concentration units of:

$$\frac{\text{mL pollutant}}{\text{m}^3\text{ mixture}} = \left(\frac{\mu\text{g pollutant}}{\text{m}^3\text{ mixture}}\right) \times \left(\frac{24.45 \times 10^{-3}}{M}\right) \tag{2.17}$$

Since 1 $m^3 = 10^6$ mL, the quantity on the left-hand side is called "the volume of pollutant per million volumes of gas" or, for short, parts per million by volume (ppmv). Often, the v is omitted, though confusion can arise with the customary liquid concentration measure of milligrams per liter, also called ppm by mass, or ppml, or just ppm. For very small concentrations, ppbv or ppb (parts per billion by volume) may be used, or even parts per trillion (ppt). Therefore, for a gaseous pollutant at 25 °C, and 1 atm

$$\text{ppmv} = \left(\mu g / m^3\right) \times \left(24.45 \times 10^{-3} / M\right)$$

or

$$\frac{\mu g}{m^3} = \frac{\text{ppmv} \times M \times 10^3}{24.45} = 40.9 \times \text{ppmv} \times M$$

$$= 40.9 \times 10^{-3} \times \text{ppbv} \times M$$

$$\text{Also: nanograms}/m^3\text{: ng}/m^3 = \text{ppbv} \times M \times 10^3/24.45$$

$$\text{nanograms/liter: ng/L} = \text{ppbv} \times M/24.45 \qquad \textbf{(2.18)}$$

Example Problem 2.4

A volume meter operates at 28 °C and an internal vacuum of –24 mm Hg. Barometric pressure is 756 mm Hg. The meter measures 13 ft^3 air. What is the volume at standard ambient conditions?

Solution

$$V_{std} = V_{meter} \times \frac{P_m}{P_{std}} \times \frac{T_{std}}{T_m} = 13 \times \frac{(756 - 24)}{760} \times \frac{298}{(28 + 273)}$$

$$= 12.4 \text{ std ft}^3$$

Example Problem 2.5

The EPA National Ambient Air Quality Standard (NAAQS) for a 1-hour average for carbon monoxide is 35 ppm. Express this in $\mu g/m^3$ and mg/m^3 at standard ambient conditions.

Solution

$\mu g/m^3$ = ppmv × M × $10^3/24.45$ = 40.9 × ppmv × M
Carbon monoxide's molecular weight is 28,
40.9 × 35 × 28 = 40,080 $\mu g/m^3$ = 40 mg/m^3.

Example Problem 2.6

40 ng/m^3 of benzene (M = 78.1) at 21 °C and 1 atm is equivalent to how many ppbv?

Solution

Note here that we do not have standard conditions; 40 ng of benzene have been measured in the following standard volume of air:

$$V_{std} = V_{meas} \times \frac{P_m}{P_{std}} \times \frac{T_{std}}{T_m} = 1\ m^3 \times \frac{1\ atm}{1\ atm} \times \frac{298}{294} = 1.014\ std\ m^3$$

Thus, the concentration at standard conditions is:

$$40/1.014 = 39.46\ ng/m^3$$

Then, $39.46 \times 24.45 \times 10^{-3}/78.1 = 0.0124\ ppbv$

In general, convert from actual measured temperature and pressure to standard conditions first; then compute ppbv or ppmv.

$$\frac{ng}{m^3 \times \left(\frac{P_m \times T_{std}}{P_{std} \times T_m}\right)} \times \frac{24.45}{M \times 10^3} = ppbv$$

or

$$\frac{\mu g}{m^3 \times \left(\frac{P_m \times T_{std}}{P_{std} \times T_m}\right)} \times \frac{24.45}{M \times 10^3} = ppmv$$

2.7 MASS CONTINUITY

From thermodynamics or fluid mechanics, the Equation of Mass Continuity is derived. For the simple case of one-dimensional flow, where all fluid flow lines can be assumed to be parallel, as in a pipe or duct (Figure 2.1), the mass rate of flow, dm/dt, is:

$$\frac{dm}{dt} = \dot{m} = \rho \times A \times v \qquad (2.19)$$

where ρ = density of the fluid
A = duct cross-sectional area
v = velocity

This mass flow rate is steady as long as there are no leaks, sources, or sinks that result in the creation, destruction, or loss of mass of the flowing fluid. Consequently,

$$\rho_1 \times A_1 \times v_1 = \rho_2 \times A_2 \times v_2 = \text{constant} \qquad (2.20)$$

where ρ_1, ρ_2 = density at location 1,2
A_1, A_2 = cross-sectional area
v_1, v_2 = velocity

This relationship, $\rho \times A \times v$ = constant (Equation 2.20), is known as the Law of Conservation of Mass and has great value in emission rate calculations.

2.8 BERNOULLI'S EQUATION

Another relationship derived from either fluid dynamics or thermodynamics of importance to air sampling is Bernoulli's Equation, a derivation from the concept of energy conservation or momentum change.

Very briefly, the First Law of Thermodynamics tells us that energy is conserved. Applied to a stream of fluid flowing in an insulated pipe or tube, heat transfer is zero, and work done on the fluid is equal to the sum of changes in internal, kinetic, and potential energy. For the simple case where temperature (and therefore internal energy) is constant and only pressure-volume work is done on the fluid, between states 1 and 2:

Net work on fluid = gain in kinetic + gain in potential energy

$$P_1V_1 - P_2V_2 = \frac{1}{2}(m_2 \times v_2^2 - m_1 \times v_1^2) + (m_2 \times gz_2 - m_1 \times gz_1) \qquad (2.21)$$

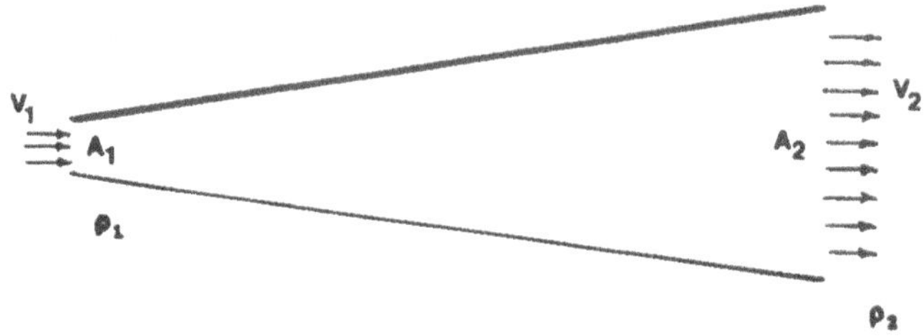

Figure 2.1. One-dimensional flow.

where V = volume
z = height
v = velocity
g = gravitational constant
m = mass

If mass is conserved, $m_2 = m_1$, and we may divide through by mass and rearrange terms:

$$P_1\upsilon_1 + \frac{v_1^2}{2} + gz_1 = P_2\upsilon_2 + \frac{v_2^2}{2} + gz_2 = \text{constant} \qquad (2.22)$$

where υ = specific volume or volume/mass.

One significant application of Bernoulli's Equation is the case when there is little or no change in altitude; $z_1 = z_2$ and specific volume can be considered a constant equal to υ. Note also that density is the inverse of specific volume; $\rho = 1/\upsilon$.

$$P_2 - P_1 = \frac{\left(v_1^2 - v_2^2\right)}{2\times\upsilon} = \frac{(v_1^2 - v_2^2)\times\rho}{2} \qquad (2.23)$$

The expression on the right of the equality, $(v_2^2 - v_1^2) \times \rho$, is called the dynamic pressure or velocity head, and Equation 2.21 allows the use of pitot tubes (which measure pressure) to measure fluid velocity, a topic explored further in Chapter 5.

CHAPTER 2 PROBLEMS

1. Convert the following temperatures:
   a. 68 °F to Celsius
   b. 10 °C to Fahrenheit
   c. 25 °C to Kelvin
   d. –25 °F to Rankine
   e. 560 °R to Kelvin
   f. 250 K to Rankine
   g. 150 °F to Kelvin

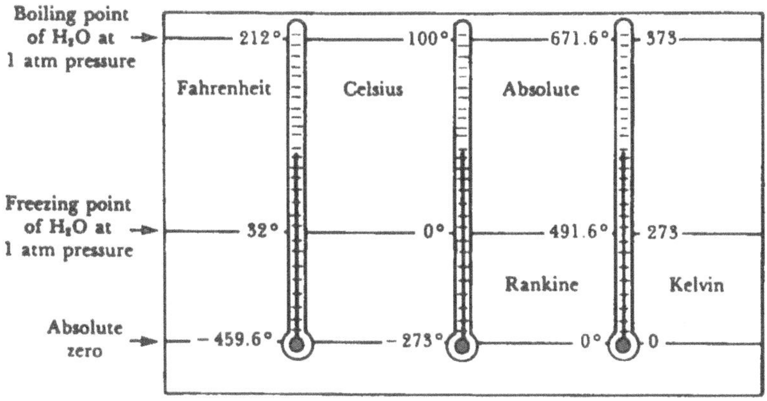

Sketch to accompany Problem 2.1

2. a. For a barometric pressure of 760 mm Hg, a system having internal pressure of 770 mm Hg would have what gage pressure?
   b. For the same barometric pressure and a gage pressure of −136 mm H_2O, what would be the absolute pressure?
3. a. The barometer scale shown below indicates a barometric pressure of how many centimeters of mercury?
   b. How many inches of mercury?

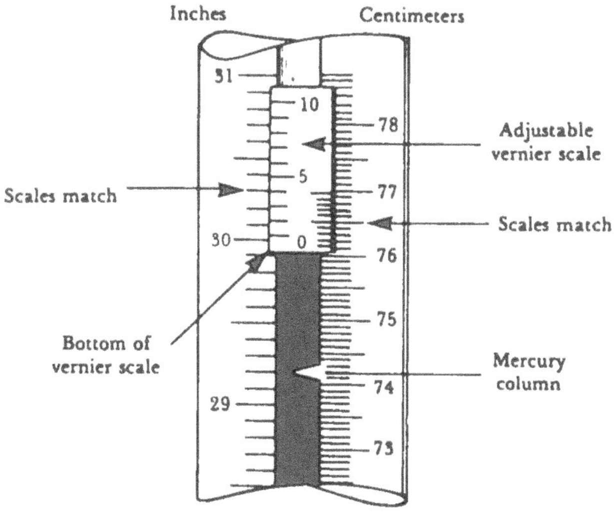

Sketch to accompany Problem 2.3

4. Convert the following pressures:
   a. 780 mm Hg to kPa
   b. 14.1 psi to " Hg
   c. 4.6" H2O to " Hg
   d. 5.6" H2O to mm Hg
   e. –3.1" H_2O gage to absolute pressure in mm Hg if P_{bar} = 1008 mB
5. 15 ft^3 of exhaust gas at 208 °F and 28.7" Hg are collected. What is the volume at the EPA stack sampling standard conditions of 20 °C (68 °F) and 29.92" Hg?
6. 2000 mL/min of air sampled for 30 min at 18 °C and 103 kPa equals what volume at standard conditions of 25 °C , and 1 atm:
   a. In liters?
   b. In m^3?
7. From the air sampled in the preceding problem, 3.2 ng toluene (M = 92.2) were collected. What is the ambient toluene concentration in ng/L and in ppbv?
8. Convert 0.12 ppm of ozone (M = 48) to µg/m^3 at standard ambient conditions.
9. Convert 240 µg/m^3 of methane (M = 16) at standard conditions to ppmv.
10. Convert 1000 µg/m^3 of SO_2 (M = 64) at standard ambient conditions to ppmv.
11. 2.1 standard liters of air were sampled through an organic vapor adsorbing medium. The following quantities of compounds were recovered:

    20.08 ng benzene (M = 78.13)
    678.91 ng toluene (M = 92.2)
    45.51 ng ethylbenzene (M = 106.2)
    86.10 ng xylene (M = 106.2)

    Give the concentration of each of these "BTEX" compounds in ppbv.
12. Ambient air was sampled at a rate of 2.25 L/min for a period of 3.25 hr at 19 °C and 748 mm Hg. What volume of air was sampled at standard conditions?
13. 13 ft^3 of air was drawn from a ventilation duct and measured by means of a dry gas meter. The meter's temperature averaged 60 °F and the meter gage pressure was –0.9" H_2O. Barometric pressure during the sampling was 742 mm Hg. Calculate the volume in standard cubic feet and in standard cubic meters (25 °C, 1 atm).
14. A dilution of benzene in methanol is made by placing 1.5 mL benzene (density of benzene is 0.879 g/mL and its molecular weight is 78.1) in a 100-mL volumetric flask and bringing to

volume with methanol. Use 3 significant digits throughout.

a. What is the concentration of benzene in the solution in mg/L, and

b. In ppml (parts per million liquid)?

15. From the solution in the preceding problem, 2 µL are drawn with a microliter syringe and injected into a gas-tight bag that contains 1 L of pure air at 25 °C and 1 atm pressure.

    What is the benzene concentration in the bag in:

    a. ppbv (parts per billion by volume)?

    b. $\mu g/m^3$ (micrograms per cubic meter)?

    c. ng/L (nanograms per liter)?

3 AIR OR GAS VOLUME MEASURING DEVICES

3.1 INTRODUCTION

The principles reviewed in Chapter 2 are utilized in this and several of the following chapters to understand the operation of many different air and gas measuring devices. There are three broad categories of such devices:

- Those that measure volume directly
- Those that measure rate of gas flow
- Those that measure velocity of gas

In this chapter, direct volume instruments will be considered; the other two categories will be the subjects of Chapters 4 and 5.

Volume meters measure the total volume, V, of gas passed through the meter while it is operated for a specific length of time. The time period, t, and also the temperature and pressure of the gas as it passes through the meter are measured along with the volume. Frequently, the average flow rate, Q, is computed from Equation 3.1.

$$Q = \frac{V}{t} \tag{3.1}$$

3.2 CALIBRATION

Before any gas measuring device can be relied upon to be accurate, it must be calibrated. The process of calibration, in simplest terms, is a comparison of the device's response to that of a device known to be correct (sometimes simply called the "known"). The frequency of calibration of any device depends on:

- Instrument characteristics — sensitivity and experience with its stability under similar use patterns
- Instrument use — rough handling, moving, heavy usage, and changing environment necessitate frequent calibration
- Instrument users — multiple users and users of various skill and experience level likely warrant more frequent instrument calibration

To calibrate a volume measuring device, a "standard meter" or one known to be correct or more accurate and precise is required, together with all ancillary equipment needed to connect, operate, and monitor both meters. There are three types of standard meters:

- Primary standard — Devices for which measuring volume can be accurately determined by measurement of internal dimensions alone; the accuracy of this type of meter is ±0.3% or more.
- Intermediate standard — Devices that are more versatile than primary standards, but for which physical dimensions cannot be easily measured. Intermediate standards are calibrated against primary standards under controlled laboratory conditions. The accuracy of this category of device is ±1 to 2%.
- Secondary standard — Devices for general use that are calibrated against primary or intermediate standards. Typically more portable, rugged, and versatile than devices in the other two categories, these devices generally have accuracies of ±5% or better.

In this chapter, the design of the following gas volume meters will be reviewed, together with a discussion of calibration techniques and applications to air sampling:

- Spirometer — primary standard
- Displacement bottle — primary standard
- Soap-bubble meter — primary standard
- Mercury-sealed piston — primary standard
- Roots meter — intermediate standard
- Wet test meter (WTM) — intermediate standard
- Dry gas meter (DGM) — intermediate standard

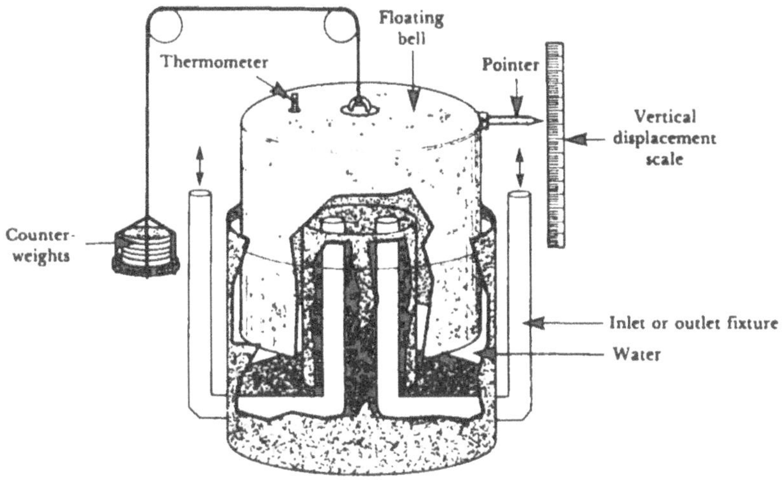

Figure 3.1. Spirometer (Reference 15).

3.3 SPIROMETER

A spirometer's primary component is a precisely manufactured cylinder with the top end closed. The volume of the cylinder can be easily measured. The open end of the cylinder is submerged in a tank of fluid (Figure 3.1). Valves can be operated to open or close the inside of the cylinder to the atmosphere. As the cylinder is lowered into the fluid (usually water), the fluid displaces air and causes it to be discharged from inside the cylinder. Counterweights allow the spirometer to operate with negligible internal gage pressure. A scale, carefully correlated with cylinder volume, allows the vertical position, h, and hence the volume of air displaced, V (Equation 3.2) to be read.

$$V = \frac{\pi \times dia^2 \times h}{4} \tag{3.2}$$

The fluid in the spirometer should be at room temperature, and care must be taken to ensure that internal pressure is allowed to equilibrate with surrounding pressure. The conditions of the measured air can then be assumed to be those in the room, and the correction of the volume measured to standard conditions is easy (see Equation 3.3).

$$V_{std} = V_{meas} \times \frac{P_{room}}{P_{std}} \times \frac{T_{std}}{T_{room}} \tag{3.3}$$

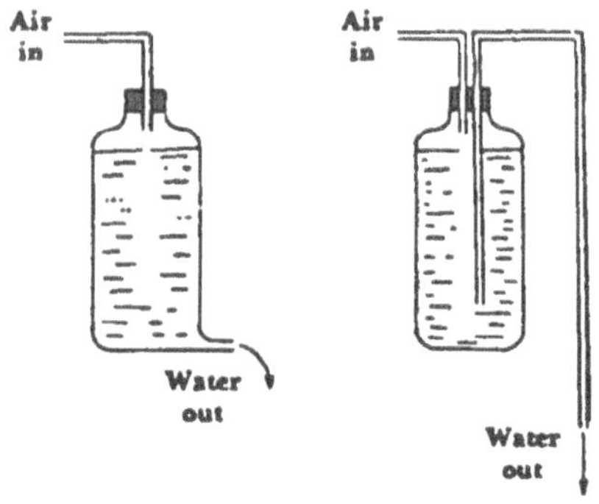

Figure 3.2. Displacement bottle (Reference 15).

Generally, evaporation of the spirometer fluid into the measured gas is not significant, and no correction is made.

Periodic calibration of a spirometer can be performed by measurement of the cylinder (called "strapping" because a measuring tape is "strapped" around the cylinder) and is described by Nelson.[8]

Clearly, the spirometer is a laboratory device that does not lend itself to portability. However, it is simple and fairly inexpensive, and often serves as a primary standard used to calibrate other devices such as dry gas meters and Roots meters.

3.4 DISPLACEMENT BOTTLE

A displacement bottle is a very simple and frequently used gas measuring device. Many other volume devices are routinely calibrated against displacement bottles. The bottle is usually a stoppered glass jar or rigid plastic container filled with water or some other liquid. The container may have a bottom drain to let water out and a hole in the stopper to let air or gas in, or it may be a siphon-type bottle with a two-holed stopper (Figure 3.2). As the liquid in the bottle is drained or siphoned out, gas is drawn in by the lowered pressure to take the place of the liquid removed. The volume of gas drawn in is equal to the volume of liquid removed, which is accurately measured with a volumetric flask or graduated cylinder. To simplify pressure and temperature corrections, the liquid should be at the same temperature as the air or gas drawn in (usually room temperature). The amount of liquid evaporation into the incoming gas is negligible.

If the displacement bottle is used to calibrate another device (see, for example, Figure 3.3), there may be substantial negative gage pressure at the bottle inlet. The pressure and temperature of the air should be measured, and the volume of gas (which is equal to the volume of liquid removed) should be corrected to standard conditions.

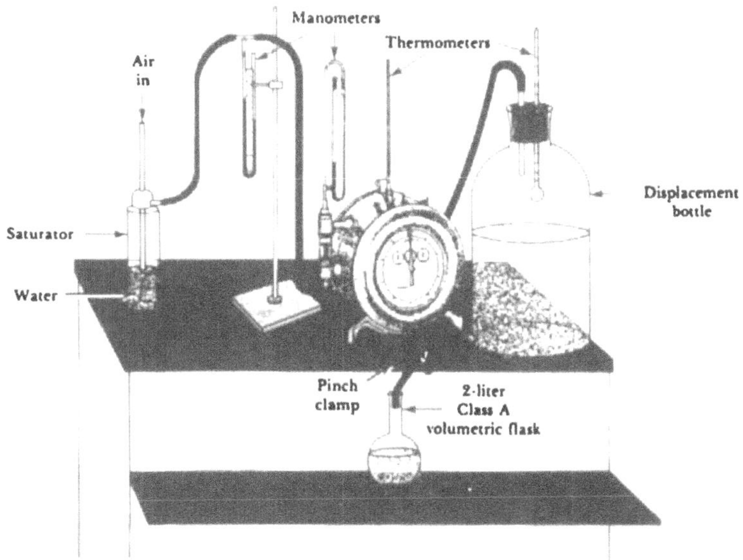

Figure 3.3. Displacement bottle used to calibrate a wet test meter (Reference 15).

3.5 BUBBLE METER

The soap-bubble meter or soap-film meter is an accurate and convenient method for measuring small volumes at gas flow rates of between 1 mL/min and 1 L/min. Bubble meters that split off and measure a fraction of the flow stream can be used for much higher flow rates. With a stopwatch, bubble meters are frequently used to measure flow rates for air samplers and gas chromatographs.

The bubble meter consists of a cylindrical glass tube with graduated markings. Several manufacturers sell bubble meters in many different diameters for various flow rate ranges. Inverted burets also make convenient bubble meters. In the simplest bubble meter, a vacuum source is connected to an inverted buret, and the open bottom is momentarily immersed in a soap and water solution. See Figure 3.4. Provided that the inside walls of the tube are wetted with the soap solution, and the upward velocity is not too great, the soap bubble will rise steadily in the tube. The volume of air swept ahead of the bubble can be measured by the graduations on the tube. Generally, the bubble's progress is timed with a stopwatch to give volume per time or flow rate. If the bubble moves too slowly or too fast (or breaks because of rapid motion), a smaller or larger diameter tube can be used. Fast-moving bubbles can cause imprecise flow rate measurements due to operator reaction time in starting and stopping the timer.

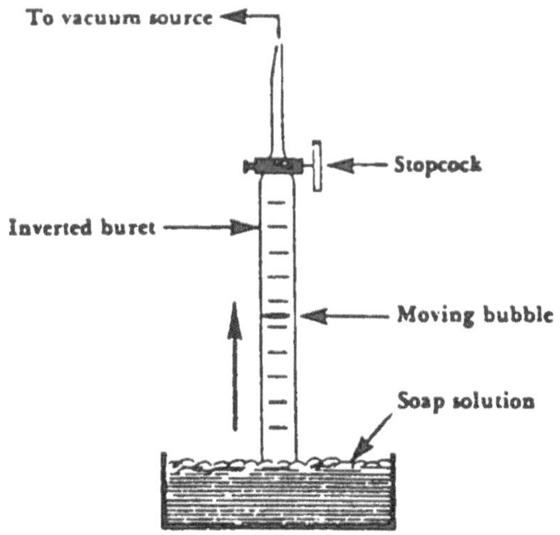

Figure 3.4. Soap-bubble meter (Reference 15).

The tube volume can be calibrated gravimetrically or volumetrically with distilled water. The bubble meter tube is inverted, filled with water, and the liquid drained from the top scale mark to the bottom into a suitable container. The volume or weight of the liquid can be measured; and with proper temperature corrections, this calibration is as accurate as the volume or weight standard.

Manufactured bubble meters can be designed to operate with either positive or negative pressure. In positive pressure bubble meters, the soap solution is contained in a rubber bulb attached to the bottom of the tube in such a way that the soap solution level can be raised just above the bottom inlet. The incoming gas forms a bubble, which then travels up the tube. See Figure 3.5.

The volume of gas swept out by the bubble must be corrected to standard conditions. The bubble provides only negligible head; thus, the gas will be measured at room temperature and pressure. However, if the gas is dry (relative humidity less than 50%), water from the soap solution will be evaporated into the gas, erroneously increasing the volume measured. By Dalton's Law of Partial Pressures, the pressure of the gas and water vapor mixture will then be equal to the pressure of the gas one wishes to measure, plus the pressure of water vapor added by the bubble meter. To correct to the original gas volume, the gas is assumed to be saturated with water vapor at its temperature. For air, tables of saturation vapor pressure as a function of temperature are available in many texts and handbooks (see Appendix G). This vapor pressure (P_{vap}) must be subtracted from the barometric pressure (P_{bar}) to give the pressure of the gas alone.

$$P_{gas} = P_{bar} - P_{vap} \quad (3.4)$$

Corrections to standard conditions using Equation 2.12 can incorporate this vapor pressure correction:

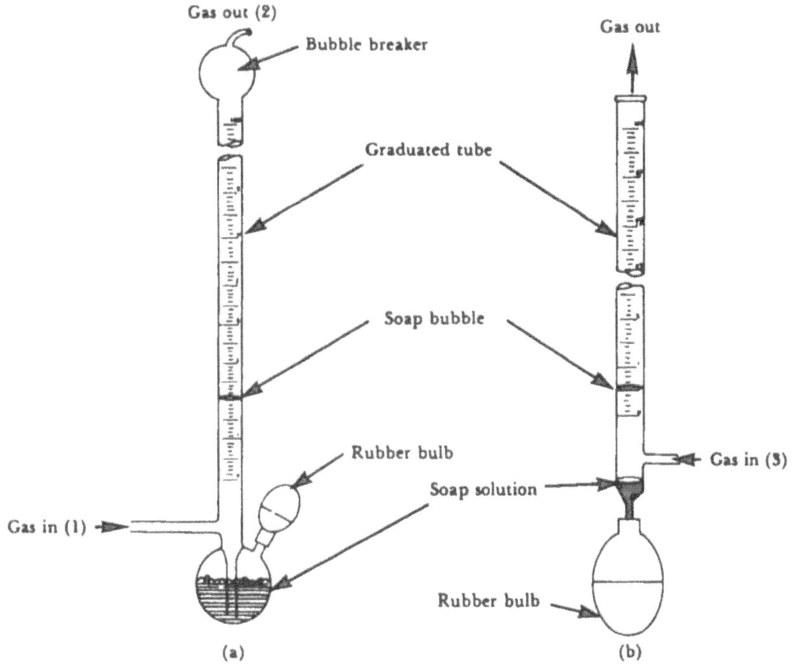

Figure 3.5. Several configurations of soap-bubble meters (Reference 15).

$$V_{std} = V_{meas} \times \frac{(P_{bar} - P_{vap})}{P_{std}} \times \frac{T_{std}}{T_{meas}} \tag{3.5}$$

A soap-bubble meter is one of the simplest primary standard devices to use, and can have accuracies of ±1% or better, depending on how accurately it is calibrated. For these reasons, bubble meters are probably the most frequently used calibration devices in air sampling. Recently developed electronic bubble meters employ an infrared light source and photocell on opposite sides of the bubble tube to measure bubble position very precisely, and a microprocessor to record time and calculate flow rate.

3.6 MERCURY-SEALED PISTON

A more accurate frictionless piston device than the bubble meter is the mercury-sealed piston. A precisely bored borosilicate glass cylinder, a close-fitting polyvinyl chloride piston, and a piston ring of mercury comprise the apparatus (Figure 3.6). The mercury stays in place around the piston because of its high viscosity and the closeness of fit between the piston and cylinder.

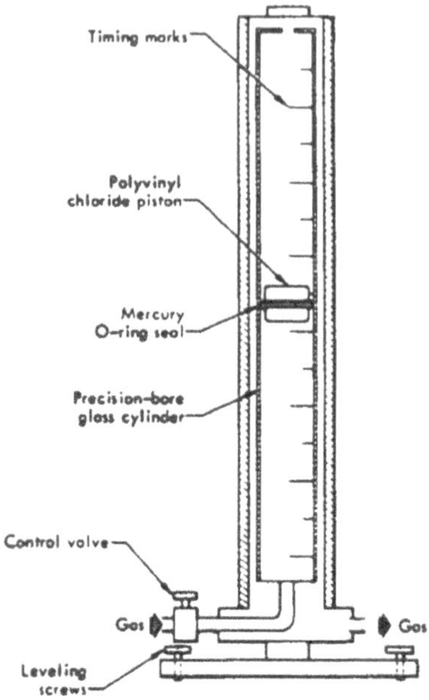

Figure 3.6. Mercury-sealed piston (Reference 8).

The instrument is virtually frictionless, but the weight of the piston must be compensated for in order to achieve the greatest accuracy. Calibration is usually performed by the manufacturer.

The mercury-sealed piston is an expensive and delicate instrument, used as a primary standard in the laboratory, and never taken to the field. The flow rate range depends on the volume of the cylinder. Models with capacities from 1 mL/min to 24 L/min are available, and accuracy is ±0.2%.

3.7 WET TEST METER

The wet test meter (WTM) is a carefully manufactured laboratory device that utilizes a water level to capture precise volumes of gas in a series of chambers attached to a rotating shaft (Figure 3.7). The meter is partially filled with water, as monitored by a sight gauge. The water level must be precisely maintained; as a means of calibration, the sight gauge indicator is adjustable. The location of the gas inlet and exit ports is such that the entering gas fills an underwater chamber, displacing the water and causing the shaft to rotate due to the buoyancy of the gas. The entrapped gas is released at the upper portion of the rotation and the chamber again fills with water. In turning, the shaft turns a pointer on the front face of the meter and, generally, a series of volume-accumulating counters that register the total volume of gas passed through the meter.[8]

Figure 3.7. Wet test meter.

As noted, the maintenance of precise water level is the key to accurate measurement, and the WTM has leveling screws, a sight gauge with settable indicator, and fill and drain cocks to allow the operator to bring the meniscus of the water to exactly the right level. Because the measured gas passes through the water, the gas will become saturated with water vapor. To minimize the loss of water by evaporation, a water impinger is often installed upstream of the WTM to presaturate the gas.

The gas will also dissolve into the water in the WTM. The gas to be measured should be allowed to pass through the WTM for 1 hr prior to measurement in order to completely saturate the water inside the meter with gas and avoid inaccuracies due to absorption of gas during measurement. The

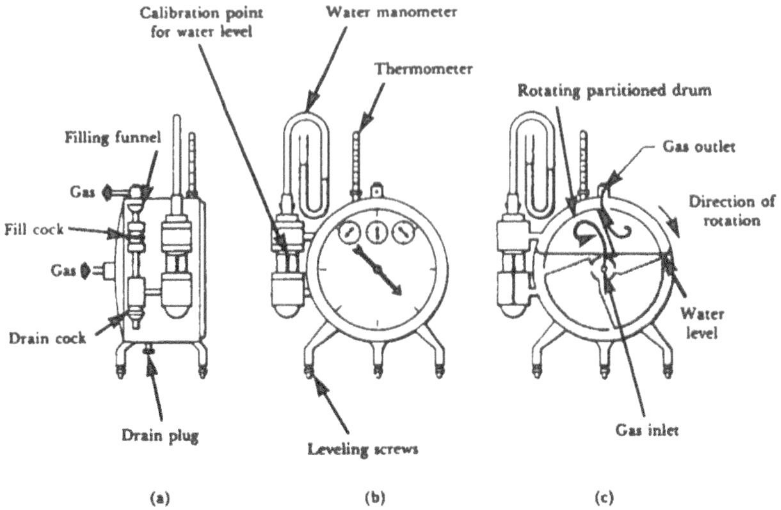

Figure 3.7a, b, c. Internal mechanism of a wet test meter (Reference 8).

water in the meter should be at the same temperature as the gas measured — usually room temperature. If water is added, time must be allowed for temperature equilibration and gas saturation. Most WTMs have a built-in thermometer and a water manometer to read internal gage pressure.

A WTM can be calibrated with a spirometer as shown in Figure 3.8. Enough gas is drawn through either system to turn the WTM at least several revolutions and significantly move the spirometer drum. The measurement should be repeated several times. To compare the spirometer and WTM volumes, absolute pressures and temperatures for both devices must be used to correct the meter readings (Equation 3.6).

$$V_{spir} = V_{wtm} \times \frac{P_{wtm}}{P_{spir}} \times \frac{T_{spir}}{T_{wtm}} \tag{3.6}$$

where the subscripts wtm = wet test meter
spir = spirometer

A displacement bottle may also be used to calibrate a WTM. The equipment is set up as shown in Figure 3.3. Prior to assembling the system, the WTM should be leveled and the water level checked. With the impinger connected, but without the displacement bottle, the meter's water should be saturated with room air by operating with a vacuum source for 1 hr. All water in the saturating impinger, the WTM, and the displacement bottle should be at room temperature. With the displacement bottle installed, the pinch clamp is opened and

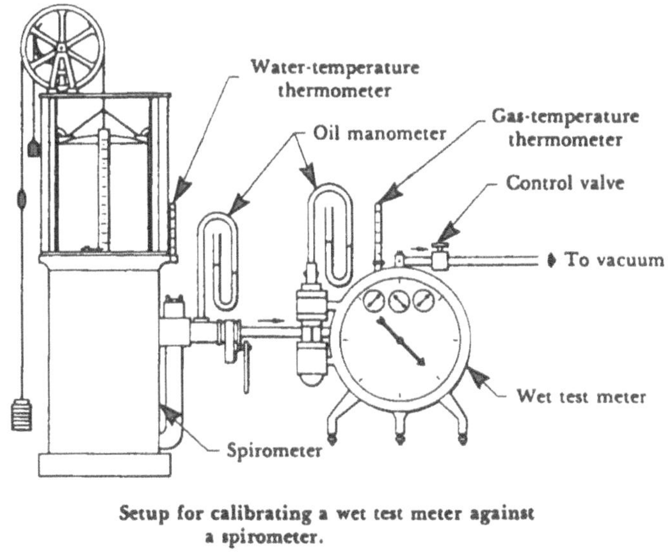

Figure 3.8. Spirometer used to calibrate a wet test meter (Reference 8).

water flow started. Without interrupting the flow, the water drain is switched to a 2-L class A volumetric flask, and the WTM readings at the start and when the flask is filled are recorded. WTM pressure and temperature are also recorded. Several repetitions are required. If the WTM measured volume is inaccurate, the sight gauge indicator may be adjusted or a correction factor, to be used with all further WTM volumes, may be calculated. Indicator adjustment, if performed, should be repeated until a correction factor, C, of less than 1 ± 0.005 is obtained.

$$C_{wtm} = \frac{V_{st}}{V_{wtm\ rdg,\ std}} \tag{3.7}$$

where V_{st} = average volume of calibration standard (displacement bottle, 2-L flask, or spirometer) for three runs, corrected to standard temperature and pressure

$V_{wtm\ rdg,\ std}$ = WTM *reading* (rdg), corrected to standard temperature and pressure

Also,

$$\%\ \text{error} = \frac{V_{wtm\ rdg,\ std} - V_{st}}{V_{st}} \tag{3.8}$$

and

$$C_{wtm} = \frac{100}{(100 + \%\ \text{error})} \tag{3.9}$$

Volumes measured by a WTM should be corrected for water vapor when measuring dry gas without a saturator (Equation 3.5). Corrections to standard conditions using the WTM temperature, gage pressure, and barometric pressure are always required.

WTMs are used as transfer standards because of their high accuracy (less than ± 1%). Because of their bulk, weight, and equilibrium requirements, they are seldom used outside a laboratory setting. Typical WTM sizes are 0.1 ft^3/revolution (or about 3 L) and 1.0 ft^3/revolution (about 30 L). WTMs can be used to measure flow rates of up to about 3 revolutions per minute. At higher rates, the meter generates too much head loss and obstructs the flow.

Example Problem 3.1

A WTM is to be calibrated with a displacement bottle. A 2-L volumetric flask is to be used. Room air at a temperature of 28 °C and pressure of 780 mm Hg will be used. The water in the displacement bottle is drained into the 2-L flask three times (See Figure 3.3). The average displacement bottle gage pressure is –25mm Hg. The average WTM reading for three trials is 0.070 ft^3, while the WTM temperature is 28 °C and gage pressure is –22 mm Hg. Compute the WTM correction factor. V_{st} = volume of air drawn into displacement bottle to replace water, corrected to standard conditions, and $V_{wtm\ rdg,\ std}$ = wet test meter reading corrected to the same standard conditions.

Solution

$V_{st} = 2 \times (780 - 25)/760 \times 298/301 = 1.967$ L
$V_{wtm\ rdg,\ std} = 0.070 * 28.32\ L/ft^3 * (780 - 22)/760 * 298/301 = 1.958$ L
$C_{wtm} = V_{st}/V_{wtm\ rdg,\ std} = 1.967/1.958 = 1.0046$

This C_{wtm} value is within the EPA limits of 1 ± 0.005, thus, the WTM is acceptably adjusted.

3.8 DRY GAS METER

Dry gas meters (DGMs) have widespread use in residential and industrial measurement of natural gas consumption, as well as in emission monitoring. They are rugged, reliable instruments that have acceptable accuracies in the

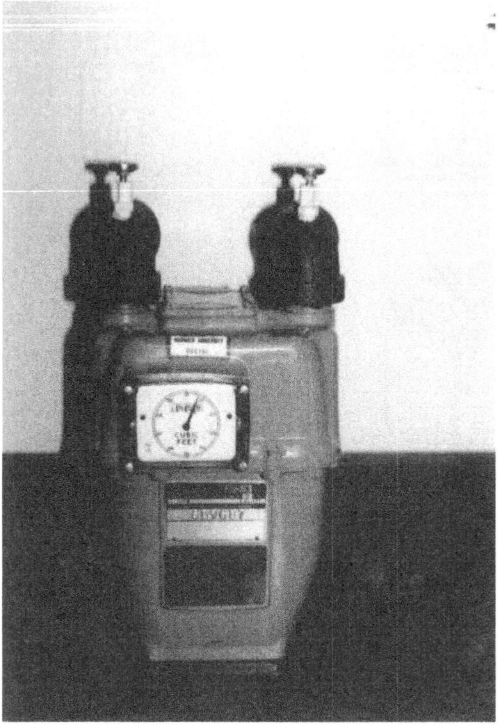

Figure 3.9. Dry gas meter.

range of ±5% and, as such, are classed as secondary standards. Dry test meters are more precise and more carefully controlled instruments than DGMs. Dry test meters employ the same metering technique as DGMs, but with a more accurate indexing method, which allows them to be used as intermediate standards. Figure 3.9 shows a standard dry gas meter.

The interior of the DGM contains two or more chambers, each containing a bellows or accordion-like inner chamber. The bellows are connected by linkages to each other and to sliding valves that direct gas flow alternately to the outside or inside of one bellows or the other, producing a reciprocating motion (Figure 3.10). The gas flow provided by either inlet pressure or outlet vacuum alternately inflates and deflates each bellows chamber, causing gas to flow through the DGM and, by means of the linkages, a set of dials to register the volume (proportional to the number of cycles).

A DGM can be calibrated against a spirometer, WTM, or displacement bottle. One big advantage of the DGM over a WTM is that no correction for water vapor is needed. The accuracy of the DGM can be corrected by adjusting the meter linkage. In practice, though, in air sampling work, if the DGM is

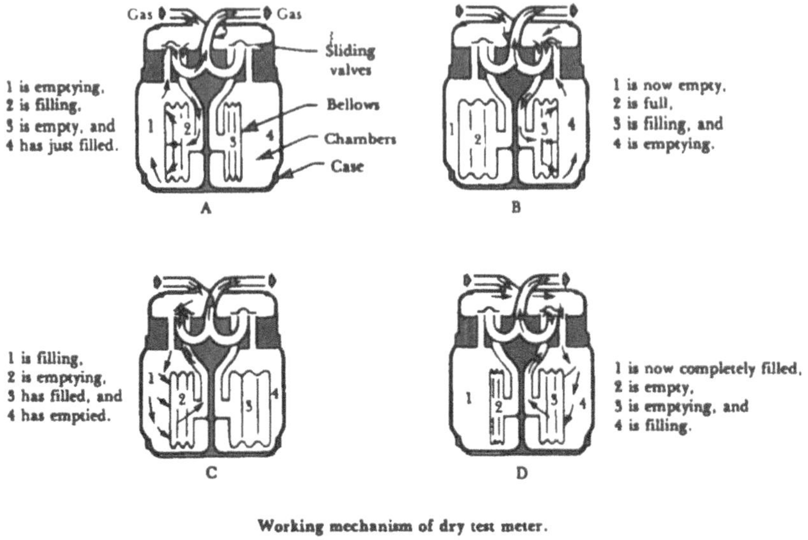

Figure 3.10. Internal mechanism of a dry gas meter (Reference 8).

within 1% of the known volume, a calibration factor is computed and used to correct all DGM values. In emission sampling work, this calibration factor, which will be discussed extensively in a subsequent chapter, is called Y, the ratio of accuracy of the DGM to its calibration device, usually a WTM.

DGMs are used extensively in air sampling field work, particularly in emission sampling. DGMs are lighter and more rugged than WTMs and much easier to transport, set up, and use. The calibration factor of a DGM (C_{dgm} or, in EPA documents, Y or Y_{avg}) must be checked frequently, however, as wear in the bellows and linkage can result in changes with use.

$$C_{dgm} = Y = \frac{V_{wtm} \times C_{wtm} \times \left(\frac{P_{wtm}}{P_{std}} \times \frac{T_{std}}{T_{wtm}} \right)}{V_{dgm} \times \left(\frac{P_{dgm}}{P_{std}} \times \frac{T_{std}}{T_{dgm}} \right)} \tag{3.10a}$$

Canceling yields:

$$C_{dgm} = \frac{V_{wtm} \times C_{wtm}}{V_{dgm}} \times \left(\frac{P_{wtm}}{P_{dgm}} \times \frac{T_{dgm}}{T_{wtm}} \right) \tag{3.10b}$$

In most EPA documentation, the following simplifying assumptions are used:

1. If $P_{wtm} \approx P_{bar}$ and $P_{dgm} = P_{bar} + \Delta H$ (where ΔH = dry gas meter head, or gage pressure) then, $C_{dgm} = Y = \{V_{wtm} \times C_{wtm}/V_{dgm}\} \times \{P_{bar}/(P_{bar} + \Delta H) \times T_{dgm}/T_{wtm}\}$
2. If temperatures are in °F, P_{bar} in " Hg, and ΔH in " H_2O, then

$$Y_i = \frac{V_{wtm} \times C_{wtm} \times P_{bar} \times \left(T_{dgm} + 460\right)}{V_{dgm} \times \left(P_{bar} + \dfrac{\Delta H}{13.6}\right) \times \left(T_{wtm} + 460\right)} \tag{3.11}$$

Equation 3.11 is the form of the DGM correction factor equation found in EPA literature. For using the DGM for emission sampling, the EPA limitation on an individual value for Y_i is is given by Equation 3.12.

$$Y_{avg} - .02 * Y_{avg} < Y_i < Y_{avg} + .02 * Y_{avg} \tag{3.12}$$

3.9 ROOTS METER

The Roots meter, also called a positive displacement meter, is a rotary meter for measuring large volumes of gas at relatively high flow rates. It is suitable for handling most types of clean, dry gases. Dust or other particulate matter in the gas measured will cause fouling of the device and excessive wear.

Roots meters consist of a chamber enclosing two precisely machined impellers, which rotate in opposite directions. The impellers have a "figure-8" cross-section and mesh together inside the casing to form a continuous tight seal (Figure 3.11). The casing is arranged with inlet and outlet gas connections on opposite sides. Impeller contours are mathematically developed and accurately produced; the correct relative impeller positions are established and maintained by precision-grade timing gears. As a result of this design, the gas inlet side of the meter is always effectively isolated from the gas at the outlet side of the impellers. Consequently, the impellers can be caused to rotate by a very small pressure drop across the meter — fractions of an inch of water.

The rotation of the impellers is in the direction indicated in Figure 3.11; as each impeller reaches a vertical position (twice in each revolution), it traps a known specific volume of gas between itself and the adjacent semicircular portion of the meter casing. Thus, in one complete revolution, the meter will measure and pass four similar volumes of gas, and this total volume is the displacement of the meter per revolution. A series of revolution counters magnetically linked to the impeller shafts read out the volume passed.

Figure 3.11. Roots meter.

The displacement volume of the Roots meter is precisely determined by the manufacturer, both by calculation and by testing it using a known volume of air or other gas. Roots meters are usually calibrated against large spirometers prior to shipment. Users do not usually have a way to calibrate Roots meters and must depend on the supplied calibration data. Volumetric accuracy of the Roots meter is permanent and nonadjustable (except for linkage adjustment) because its measuring characteristics are established by the dimensions and machined contours of nonwearing fixed and rotating parts.

The revolutions of the impellers are indexed with the meter reading calibrated in a volume unit (cubic feet or meters). Units are available that have temperature compensation devices, but corrections to standard temperature and pressure conditions are easily made. The gage pressure at both the inlet and outlet of the meter must be measured in order to include the pressure drop across the meter, ΔP, in the volume correction Equation 3.13.

$$V_{std} = V_{meter} \times \frac{(P_{inlet} - \Delta P)}{P_{std}} \times \frac{T_{std}}{T_{meter}} \qquad (3.13)$$

The primary application of Roots meters in air sampling has been as a flow calibration standard for *total suspended particulate* samplers (HiVols), PM10 and PUF samplers, and other high-flow sampling devices.

CHAPTER 3 PROBLEMS

1. The WTM of Example Problem 3.1 is used to calibrate a DGM. The WTM is connected in series with a pump, and then the DGM. The pump draws room air through the WTM and exhausts it to the DGM. An impinger is connected to the inlet of the WTM and a flow-restricting orifice (which provides a head loss, ΔH) to the exit of the DGM. Given the following data and figure, compute the DGM calibration factor.

   P_{bar} = 29.85" Hg
   $P_{g,dgm}$ = ΔH = 1.5" H_2O
   $P_{g,wtm}$ = −0.25" H_2O
   T_{dgm} = 67 °F
   T_{wtm} = 65 °F
   WTM reading = 143.6 L
   DGM reading = 150 L

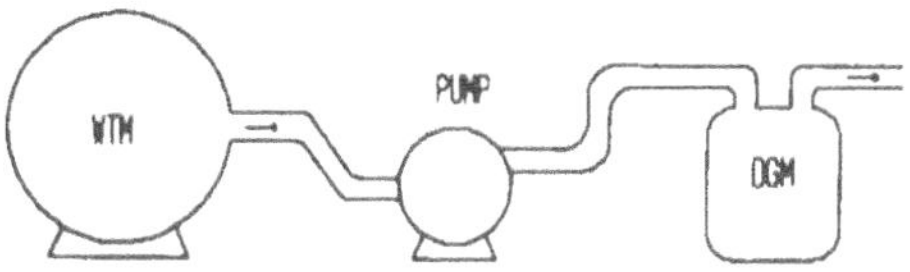

2. A 100-mL buret is to be calibrated for use as a soap-bubble meter. The buret is filled to the 100-mL mark with distilled water and drained into a tared flask three times.

   T_{water} = 23 °C (at this temperature, water density = 0.99756 g/mL)
   Tared flask weight = 73.905 g
   Filled flask weights
   1: 174.211 g
   2: 173.875 g
   3: 173.920 g

   Find the average percent error for the bubble meter volume.
3. The bubble meter of Problem 2 is used to check an ambient sampling pump's flow rate. The bubble rises through 100 mL in 60 sec. The air in the room has T = 22 °C, P = 100.55 kPa, and relative humidity = 20%. What volume does the pump handle in 60 sec, corrected to standard temperature and pressure?

4. Using a siphon system, exactly 4.5 L air were drawn through a WTM; the WTM dial reading and the pressure and temperature during the three trials were:

P_w (" H_2O)	–3	–4	–7
T_w (°C)	21	22	21
V_w (L)	4.69	4.63	4.86

P_{atm} = 29.5" Hg T_{atm} = 21 °C

a) Compute C_{wtm}, the WTM correction factor.

b) When the WTM indicated 4.56 L in a later project, and for T_{wtm} = 21 °C, P_{wtm} = 3" H_2O, P_{bar} = 29.5" Hg, what was the equivalent correct measured air volume at 25 °C and 29.92" Hg?

5. A WTM with C_{wtm} = 0.989 is used to calibrate a DGM. The equipment is set up as in Problem 1. Data are:

T_{dgm} = 77 °F
T_{wtm} = 72 °F
P_{bar} = 28.7" Hg

	V_{wtm} (ft^3)	V_{dgm} (ft^3)
DGM ΔH (" H_2O)	**(as read from meter)**	
1. 0.5	3.7	3.5
2. 1.0	4.6	4.5
3. 3.0	4.9	4.7

Does the DGM meet the EPA tolerance requirement?

6. The DGM of Problem 5 is later used and the following data recorded:

T_{dgm} = 90 °F
P_{bar} = 30.0" Hg
ΔH = 3" H_2O
DGM reading = 5 ft^3

Calculate the volume passed through the DGM at standard temperature (77 °F) and pressure.

4 GAS FLOW RATE MEASURING DEVICES

4.1 INTRODUCTION

It is often desirable to the measure rate of flow of a gas. Even if the volume of the gas is the desired quantity, rate and time are often measured and multiplied together to compute volume. A good example is in PM10 ambient sampling, where flow rate for 24 hr is used to find total volume of air sampled.

Flow rate measuring devices utilize some property of the fluid to be measured — generally, density or viscosity — as well as gas laws or fundamental thermodynamic properties of the fluid. Devices considered in this chapter are the orifice meter, venturi meter, rotameter, and mass flow meter.

4.2 CALIBRATION

All the comments in Chapter 3 relative to calibration are as pertinent to flow rate measuring devices as they are to volume devices. Before use, any flow rate measuring device must be calibrated. The process involves connecting the meter in series with a volume or flow rate instrument known to be correct, making Ideal Gas Law adjustment for temperature, pressure, and humidity differences between the two devices and developing a correction factor or "calibration curve" for the the meter being calibrated. Correct and accurate use of any flow rate measuring device requires knowledge of general principles of

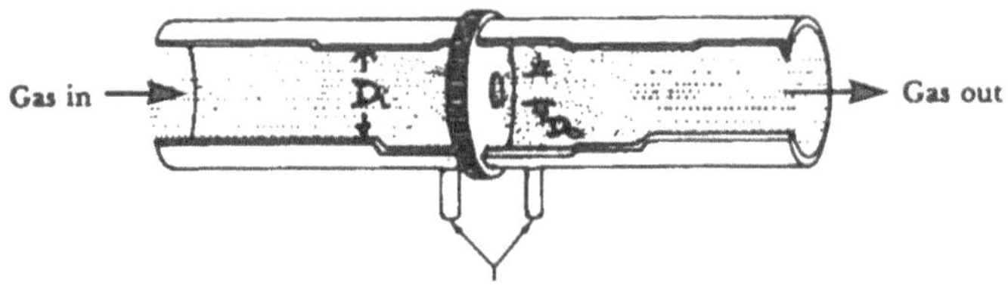

Figure 4.1. Sharp-edged orifice meter (Reference 15).

thermodynamics and fluid flow, and an understanding of how they are employed by the particular instrument.

All the flow rate measuring devices reviewed in this chapter are considered *secondary standards* — they must be calibrated against a primary or intermediate standard volume measuring device, together with an accurate stopwatch.

4.3 ORIFICE METER

An orifice meter, or variable pressure meter, is a commonly used device in air sampling. The flow rate is a function of the pressure drop across a constriction (orifice) in the flow stream and the size of the orifice.

The simplest orifice consists of a thin plate with one sharp-edged circular hole in the center inserted into a straight section of pipe or duct. See Figure 4.1. Pressure taps on either side of the orifice allow upstream and downstream pressure, or the pressure difference across the orifice, to be measured with a gauge or manometer. Two applications of orifices to air sampling work are flow rate monitoring in the EPA Method 4 and 5 Sampling Trains, where a 1/4" orifice in a 1/2" pipe is used (Figure 4.2) and in the calibration unit historically used for high-volume samplers and some PM10 samplers, where 1/2" to 1" orifices are used (Figure 4.3).

The pressure drop across orifices in air sampling work is typically in the range of 0 to 20" of water; thus water manometers or equivalent pressure gauges are used.

From the First Law of Thermodynamics for an open system, the fundamental relationship between pressure and flow rate can be derived.

$$Q_H + U_i + \left[P_i \times \upsilon_i + g \times z_i + \frac{1}{2} \times v_i^2\right] \times m_i =$$

$$U_o + \left[P_o \times \upsilon_o + g \times z_o + \frac{1}{2} \times v_o^2\right] \times m_o + W \qquad \textbf{(4.1a)}$$

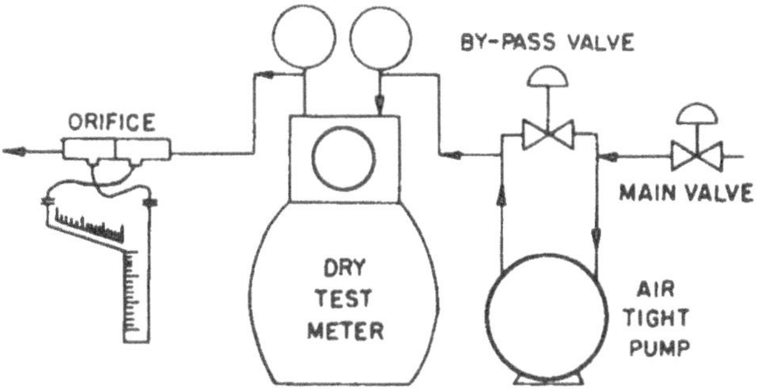

Figure 4.2. Orifice in use.

where Q_H = heat added to the fluid
U = internal energy of the fluid
P = pressure
υ = specific volume
g = gravitational constant
z = elevation
v = velocity
m = mass of fluid
W = work done by the fluid
The subscripts *i* and *o* refer to inlet and outlet of the orifice, respectively.

If we assume that mass is conserved and flow of gas through the orifice takes place adiabatically (i.e., no heat flow; $Q_H = 0$) and with no internal energy change of the gas ($U_i = U_o$), work done by the gas ($W = 0$), or change in height ($z_i = z_o$), the First Law simplifies to Equation 4.1b.

$$\left(P_i \times \upsilon_i\right) + \left(\frac{1}{2} \times v_i^2\right) = \left(P_o \times \upsilon_o\right) + \left(\frac{1}{2} \times v_o^2\right) \qquad \textbf{(4.1b)}$$

where υ_i, υ_o are specific volumes of gas at orifice inlet and outlet
v_i, v_o are velocities
P_i, P_o are pressures

The gas flowing through the device can be assumed incompressible ($\upsilon_i = \upsilon_o = \upsilon$ = constant) if the pressure drop across the orifice is small compared to the upstream pressure. With this assumption, and noting that density, ρ, is the inverse of specific volume, Equation 4.1b becomes Equation 4.2.

$$\left(P_i - P_o\right) * \upsilon = \frac{\left(P_i - P_o\right)}{\rho} = \frac{1}{2} * \left(v_o^2 - v_i^2\right) \qquad \textbf{(4.2)}$$

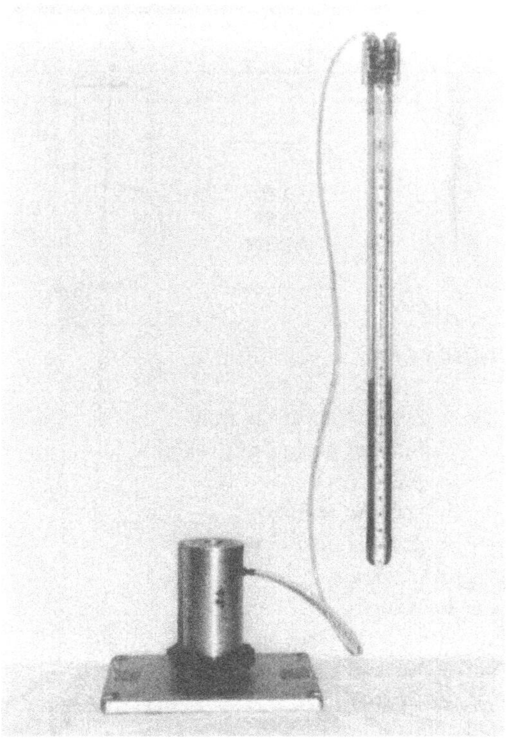

Figure 4.3. High-volume sampler orifice.

For the small pressure drop of orifices used in air sampling, volume flow rate, Q, through the orifice is equal to $A_o v_o$, where A_o is the orifice area. By continuity, the inlet velocity, v_i, is equal to the velocity through the orifice times the square of the ratio of orifice and upstream diameters (again assuming incompressibility) (Equation 4.3).

$$v_i = v_o \times \left(\frac{D_o}{D_i}\right)^2 \qquad (4.3)$$

Furthermore, the flow through an orifice is not perfectly frictionless; thus, a coefficient K that depends on sharpness and smoothness of the orifice (on the order of 0.6) is incorporated into the equation. Substituting into Equation 4.2 and rearranging terms results in Equation 4.4,

$$Q = v_o \times A_o = K \times A_o \times \left[2 \times \frac{(P_i - P_o)}{\rho}\right]^{\frac{1}{2}} \qquad (4.4)$$

where the constant K incorporates both the orifice coefficient and the diameter ratio (D_o/D_i). The orifice coefficient, and hence, K is assumed constant over all flows, though it in fact varies slightly and is a function of Reynold's Number.[32,33] Since ideal gas behavior can usually be assumed, Equation 2.6 gives:

$$\frac{P}{\rho} = R_u \times \frac{T}{M}$$

and for the orifice inlet conditions (i):

$$\frac{1}{\rho} = R_u \times \frac{T_i}{P_i M}$$

Thus, Equation 4.4 becomes

$$Q = K \times A_o \times \left[2 \times (P_i - P_o) \times R_u \times \frac{T_i}{P_i \times M}\right]^{\frac{1}{2}}$$

$$= K \times A_o \times \left[2 \times R_u \times \frac{T_i}{P_i \times M}\right]^{\frac{1}{2}} \times (\Delta P)^{\frac{1}{2}}$$

$$= K_m \times \left[\frac{T_i}{P_i \times M} \times (\Delta P)\right]^{\frac{1}{2}} \qquad \textbf{(4.5a)}$$

where K_m incorporates all the constant terms and has (SI) units of $m^2X\ [(m^3\text{–}kPa)/(kgmole\text{–}K)]^{1/2}$. If T_i, P_i, M and ΔP have standard SI units, the volume flow rate, Q, has units of m^3/s. Similar care with U.S. Customary units results in flow rate units of ft^3/s.

Or, using the term ΔH for the head loss across the orifice in place of ΔP, the standard orifice equation becomes:

$$Q = K_m \times \left[\frac{T_i}{P_i \times M} \times (\Delta H)\right]^{\frac{1}{2}} \qquad \textbf{(4.5b)}$$

Recall that throughout this derivation, incompressibility has been assumed, requiring "small" ΔP. If "small" ΔP is defined as less than $^1/_{10}$ of P_i, then for orifices used in air sampling, where P_i is approximately 1 atm, ΔP is limited to about 40" H_2O. Some investigators suggest a rule of thumb that maximum ΔP in *inches of water* be less than P_i in *psia*. This guideline would limit the maximum ΔP to about 15" H_2O for atmospheric sampling.[8]

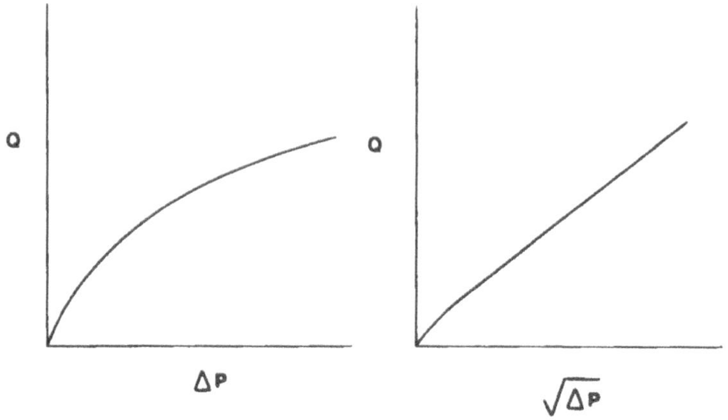

Figure 4.4. Orifice calibration curve.

In practice, no attempt is made to measure orifice dimensions or compute the coefficient. Instead, an orifice calibration curve is developed for a selected set of calibration inlet temperature, T_c, pressure, P_c, and molecular weight, M_c, by calibrating against a primary or intermediate standard device. See Figure 4.4.

A calibration curve prepared in this fashion should be clearly labeled with the orifice operating conditions during calibration (P_c, T_c, M_c). The values for Q are taken from the calibrating device and may be in units of standard volumes per unit time if Equation 2.12 is used to adjust the actual volume recorded. If the orifice is used at some other conditions (P_a, T_a, M_a), the value read from the above calibration curve is called the "indicated" standard flow rate, Q_i at that ΔP. The "actual" standard flow rate is different due to the incorporation of P_a, T_a, and M_a in the basic orifice equation, Equation 4.5. To find this actual standard flow rate, use Equation 4.6.[5]

$$Q_{act} = Q_i \times \left(\frac{P_c}{P_a} \times \frac{T_a}{T_c} \times \frac{M_c}{M_a} \right)^{\frac{1}{2}} \qquad (4.6)$$

It is important to understand the difference between use of Boyle's and Charles' Laws (see Equation 2.12) to adjust a recorded volume or flow rate to the equivalent at standard temperature and pressure, and the operation of Equation 4.6, which corrects for *operating* the orifice at noncalibration conditions.

Example Problem 4.1

An orifice meter was calibrated by the manufacturer for air at 25 °C and 1 atm. It is installed in an air sampling system where it operates at 33 °C, 0.9 atm,

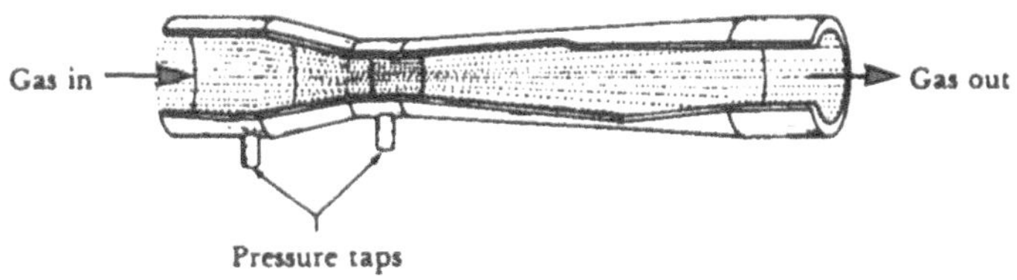

Figure 4.5. Venturi meter (Reference 15).

and reads a ΔP of 4.8" H_2O. The calibration curve (like Figure 4.4) has the relationship $Q_i = 4.1 \times (\Delta P)^{1/2}$ in slpm (liters per minute at 25 °C, 1 atm). What is the true gas flow rate in slpm?

Solution

Since the orifice is to be *operated* at conditions different than calibrated, Equation 4.6 is appropriate; Equation 2.12 is not needed here. From the calibration curve, $Q_i = 4.1 \times (4.8)^{1/2} = 8.98$ slpm.

$$Q_{true,\ std} = Q_i \times \left[\frac{33 + 273}{25 + 273} \times \frac{1}{0.9} \times \frac{28.96}{28.96}\right]^{\frac{1}{2}} = 9.59 \text{ slpm}$$

4.4 VENTURI METER

A venturi meter is a device which consists of a variable-diameter flow path — a convergent inlet section, a constant diameter throat section, and a divergent outlet section. As in an orifice meter, two pressure taps provide the flow measurement parameter. In the venturi, one tap is upstream of the inlet section, the other in the throat section. As the gas flows through the venturi, the velocity increases and pressure decreases, reaching a minimum in the throat (see Figure 4.5).

The principle of operation of the venturi is similar to that of the orifice meter — the flow rate is proportional to the square root of the pressure drop. Equations 4.4 through 4.6 apply to venturi meters, and venturi meters are calibrated in a manner identical to orifice meters. As with orifice meters, it is important to remember two limitations:

1. The coefficient employed to account for friction is assumed constant for the entire operating range
2. Flow is assumed incompressible

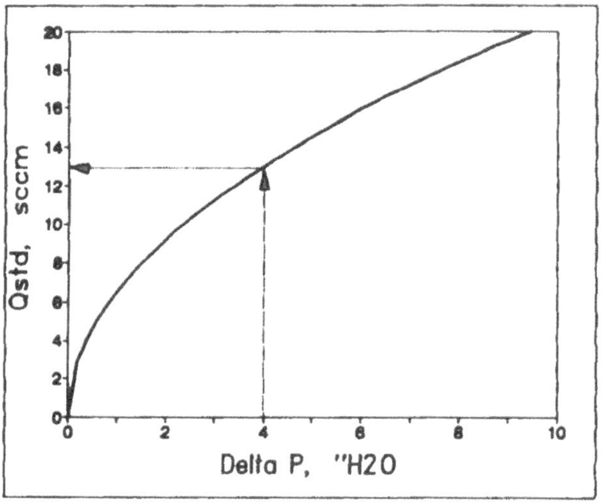

Figure 4.6. Venturi curve for example.

Together, these limitations require that the venturi meter be operated only over a narrow range of pressure drops in the proximity of those used during calibration.

Example Problem 4.2

A venturi meter was calibrated with air at 60 °C and 1.5 atm. It is to be used to measure the flow of nitrogen at 20 °C and 0.9 atm. A calibration curve is shown in Figure 4.6. With nitrogen at the use conditions, a pressure drop of 4" H_2O is measured. What is the true nitrogen flow rate in standard cubic meters per minute (scmm)?

Solution

From the calibration curve, the indicated flow rate at 4" H_2O is 13 scmm. The actual standard flow rate, using Equation 4.6, is

$$Q_{true,\,std} = 13 \times \left(\frac{293}{333} \times \frac{1.5}{0.9} \times \frac{28.96}{28.0} \right)^{\frac{1}{2}}$$

4.5 CAPILLARY TUBE

A variation on the sharp-edged orifice meter is the capillary tube (see Figure 4.7). Capillary action involves the drawing of fluid along the walls of a slender

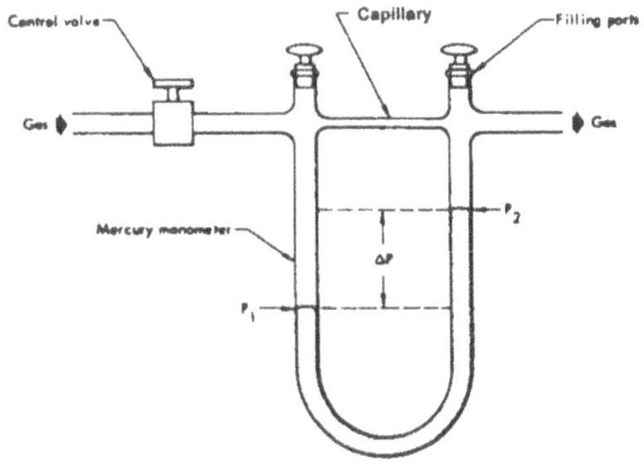

Figure 4.7. Capillary tube meter.

tube as a function of fluid properties, tube diameter and material, length, and propelling force. A slender tube with a presssure drop ($P_{up} - P_{dn}$) has a flow rate that follows Poiseuille's Law (Equation 4.7).

$$Q = \frac{\left(P_{up} - P_{dn}\right) \times \left\{\frac{\left[1 + \left(P_{up} - P_{dn}\right)\right]}{2 \times P_{dn}}\right\}}{r} \tag{4.7}$$

where Q = flow rate
P_{up}, P_{dn} = upstream and downstream pressures, respectively
r = capillary flow resistance, a function of length, diameter, and fluid viscosity

As with orifice meters, an empirical relationship between flow rate and pressure drop is developed by calibration. Simplistically, capillaries can be treated like orifices and venturies. As Equation 4.7 implies, for small pressure drop compared to P_{dn}, the flow rate is nearly linearly proportional to pressure drop (the last term in the numerator is negligible).

Capillary tubes are widely used for small flow rates (i.e., less than 100 sccm).[8] They are often built into instruments to maintain constant flow rates where pressure, temperature, and viscosity will not change. A single point calibration is sufficient in such an application. A disadvantage of capillaries is that any particle build-up can change the effective diameter and change the flow rate.

4.6 ROTAMETER

A rotameter, or visi-float or variable area meter consists of a glass or plastic tube with a steadily tapered bore and a "float" or ball of diameter slightly larger

Figure 4.8. Rotameters.

than the bore's minimum (see Figure 4.8). Many modern rotameters use a spherical float or a smooth float of a more complex design, but early rotameter floats had spiral grooves to enhance rotation and centering; hence, the name.

The tube must be kept vertical; as the fluid flows from bottom to top, the float rises. The annular area between the tube wall and the float acts as an orifice. The orifice equation (Equation 4.5) holds:

$$Q = K \times A_o \times \left(\frac{2 \times R_u \times T_i}{P_i \times M} \right)^{\frac{1}{2}} \times \left(P_i - P_o\right)^{\frac{1}{2}} \qquad (4.5)$$

where A_o is a variable annular area around the float that is directly proportional to the height of the float in the tube. Because the tube is marked with a graduated scale, the vertical position of the widest part of the float can be considered a "rotameter reading" (RR). The float rises until the upward force of viscous drag balances the downward force of gravity less buoyancy. That is, $\Sigma F_{up} = \Sigma F_{down}$ or $F_{drag} = F_{grav} - F_{buoy}$, such that

$$C_D \times A_F \times \rho \times \frac{v^2}{2} = \rho_F \times g \times V_F - \rho_g \times g \times V_F \qquad (4.8)$$

where C_D = float drag coefficient (a function of Reynold's Number)
A_F = cross-sectional area of float
ρ_F, ρ_g = density of float and gas, respectively
V_F = volume of float
v = gas velocity

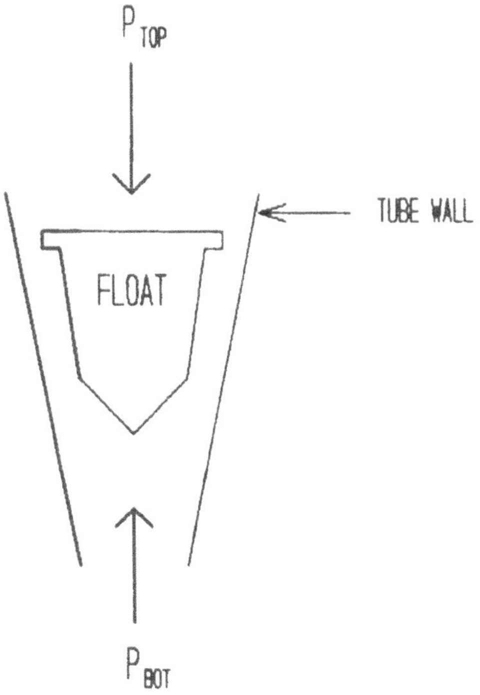

Figure 4.9. Upward force on rotameter float.

In the rotameter (see Figure 4.9), the upward force can also be stated as $(P_{bot} - P_{top}) \times A_F = \Delta P \times A_F$.
Most rotameters operate in the turbulent flow regime; and for turbulent flow around the float, the drag force is proportional to this upward "pressure force", and thus proportional to the net downward forces. For gases, the buoyancy term in Equation 4.8 is negligible, so that:

$$\Delta P \times A_F \propto \rho_F \times g \times V_F \tag{4.9}$$

In this equation, no terms appear for gas velocity or float height, indicating that the pressure drop across the float is constant. This observation is the essence of rotameter theory — the tube taper is such that the pressure drop across the annular orifice between wall and float is constant, but the *area* of the orifice changes with flow rate. For further insight into this derivation of rotameter equations, see Achinger's work.[15] Since the rotameter acts as an orifice, Equations 4.5 and 4.9 may be combined to give Equation 4.10.

$$Q = K' \times A_o \times \left(2 \times R_u \times \frac{(T_i)}{P_i \times M} \times \frac{\rho_F \times g \times V_F}{A_F}\right)^{\frac{1}{2}} \tag{4.10}$$

Since the coefficient K′ incorporates flow, meter, and gas characteristics, it is a function of Reynold's Number. K′ must be found empirically and is approximately constant only over a limited range. However, within this range, the only nonconstant term on the right-hand side of Equation 4.10 is the annular orifice area around the float, A_o; thus, the flow rate is linearly proportional to this area (and thus proportional to the height or rotameter reading, RR). See Equation 4.11.

$$Q = K_R \times RR \tag{4.11}$$

Generally, the manufacturer performs a calibration at some specified temperature, pressure, and molecular weight and marks the rotameter scale with RR values accordingly. Usually, the scale is linear, as suggested by Equation 4.11. However, if the rotameter is installed where it operates at a temperature, pressure, and/or molecular weight gas different from those used for initial calibration, the scale readings or the manufacturer's calibration curve are not applicable. See Figure 4.10.

As Equation 4.10 indicates, flow rate is a function of the square root of rotameter operating temperature, pressure, and molecular weight in the same fashion as orifice and venturi meters. Consequently, Equation 4.6 applies here as well.

Two precautions should be noted.

1. Since K_R is a function of Reynold's Number, conditions of use for a rotameter should not vary greatly from calibration conditions; in fact, *in situ* calibration of the meter is generally the best idea.
2. Rotameters often have a needle valve at the bottom, — just upstream (occasionally, one is installed at the top instead). There can be a significant pressure drop across the needle valve, such that upstream pressure is a good deal higher than the actual operating pressure inside the rotameter. With a top valve rotameter, pressure inside is higher than downstream pressure. Again, *in situ* calibration obviates any concerns over pressure differences between calibration and use. Calibration can be done with any primary or intermediate standard — wet test meter, soap-bubble meter, etc.

Rotameters are frequently built into air monitoring or other gas handling instruments to provide a quick and easy way to check flow rate. They are by far the simplest and most easily read flow measuring device. Properly calibrated, errors of less than 5% can be expected.

Example Problem 4.3

A rotameter was calibrated with air at 18 °C and 29.92" Hg. The rotameter was then used to monitor the flow of helium into a gas chromatograph at 18 °C

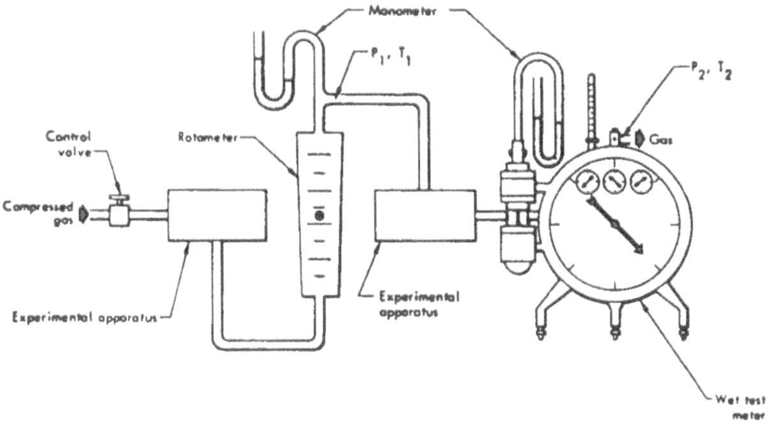

Figure 4.10. Calibration of rotameter (Reference 8).

and 29.92" Hg. The rotameter reading was 28.3 "scale units" for the helium flow. The calibration literature that the manufacturer provided with the rotameter indicates that this reading means 28.3 actual milliliters per minute (amlpm) of air at the calibration conditions. What is the actual flow rate of helium at the measured conditions?

Solution

Here, no pressure or temperature correction is required.

$$Q_{He} = Q_{air} \times \left(\frac{M_{air}}{M_{He}}\right)^{\frac{1}{2}} = 76.2 \text{ amlpm}$$

Example Problem 4.4

A rotameter reads 4.13 units when air at 745 mm Hg and 27 °C flows through it. A WTM in series gives

$$Q_{wtm,\, std} = \frac{V_{wtm} \times C_{wtm}}{\Theta} \times \frac{P_{wtm}}{P_{std}} \times \frac{T_{std}}{T_{wtm}} = 2.0 \text{ slpm}$$

This equation is an adaptation of Equation 2.12 for conversion to standard conditions, where V_{wtm} is the WTM reading for this calibration run, C_{wtm} is the WTM calibration factor found in a previous calibration, and Θ is time.

a) For use of the rotameter on air at 560 mm Hg and 100 °C, what is the flow rate corrected to standard conditions when the rotameter reads 4.13?

Solution

$$Q_{use,std} = Q_i \times (P_{cal}/P_{use} \times T_{use}/T_{cal})^{1/2}$$

where Q_i is the "indicated" flow rate from the calibration information at the RR of 4.13; i.e., 2.0 slpm.

$$Q_{use,\,std} = 2.0 \times (745/560 \times 373/300)^{\frac{1}{2}} = 2.57 \text{ slpm.}$$

b) Now, suppose it is desired to set this rotameter to allow, instead, a "use" flow rate of 2.0 slpm when operating at 560 mm Hg and 100 °C.

Solution

A calibration curve is needed for either:

i) slpm when operating at 560 mm Hg, 100 °C (this would be easiest; flow rate could be read directly from the curve) or for

ii) slpm at the original calibration conditions of 745 mm Hg and 27 °C.

Selecting the second option (the calibration plot is shown in Figure 4.11), $Q_{use,\,std}$ is desired to be 2.0 slpm:

$$Q_{use,\,std} = 2.0 = Q_i \times (P_{cal}/P_{use} \times T_{use}/T_{cal})^{\frac{1}{2}}$$

Here, Q_i is to be calculated and is the standard flow rate at *calibration* conditions; that is, the value which would be used to enter the calibration curve to find the proper rotameter reading (or read directly from the rotameter scale if the roatmeter is so marked) that corresponds to 2.0 slpm at *use* conditions. Solving, $Q_i = 1.56$ slpm. Again, this is the flow rate from the calibration curve that gives the same RR as 2.0 slpm at *use* conditions. The calibration curve is shown in Figure 4.11, from which one reads the desired RR of 3.4 to achieve the objective of 2.0 slpm when the rotameter operates at 560 mm Hg and 100 °C.

4.7 MASS FLOW METER

A device that is becoming much more widely used in the air sampling field is the electronic flow meter. Electronic mass flowmeters operate on the principle that when a gas passes over a heated surface, heat is transferred from the surface to the gas. The rate of heat transfer is related to the temperature difference, area, and heat transfer properties of the surface and gas. For gases that approximate ideal behavior, the quantity of heat, Q_H, that causes a specified temperature change is related to mass and a unique heat capacity or

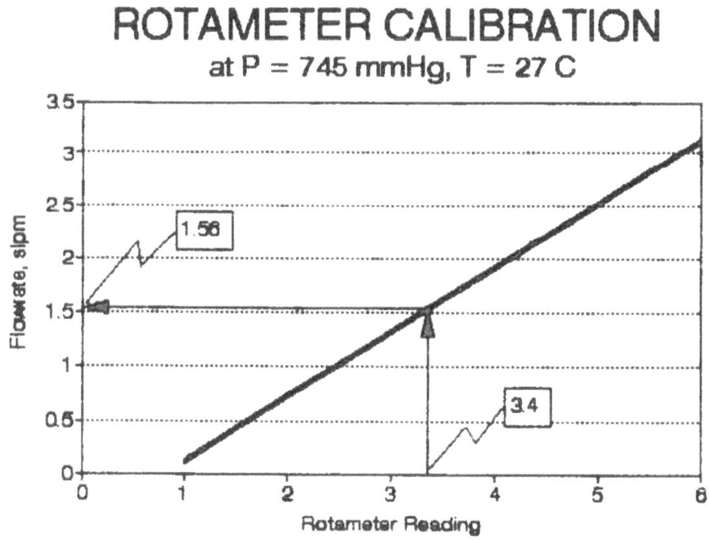

Figure 4.11. Rotameter curve for Example 4.4.

specific heat, C_p, for that gas. Over relatively small temperature changes, the specific heat is nearly constant, and the relationship is:

$$Q_H = m \times C_p \times (T_2 - T_1) \tag{4.12}$$

The *rate* of heat flow into a gas is therefore proportional to the *rate* of mass flow for a given temperature change.

There are two basic designs of mass flow meter — heated tube and insertion types (see Figure 4.12). In the heated tube type, gas flows through a tube where it is first heated by the hot walls of the tube; then the warm gas gives up its heat to a cooler section of the tube. The temperatures in the two sections of tubing are monitored with a Wheatstone bridge circuit. For a gas of known C_p, the temperature changes and the fixed rate of heat transfer to the gas allow direct calculation of the mass flow rate by Equation 4.13.

$$\dot{Q}_{\dot{H}} = \dot{m} \times C_p \times (T_2 - T_1) \tag{4.13}$$

In the insertion-style flow meter, two slender probes are used to measure temperature and thus gas flow rate. One probe measures gas temperature, while the other is heated to a specific temperature above the first. The amount of energy needed to keep the second probe at the correct temperature is proportional to the heat transfer rate to the gas and thus to the gas mass flow rate.

Since the heat capacity of each gas is unique, the mass flow meter must be calibrated for each gas to be measured, or a specific heat correction factor employed.

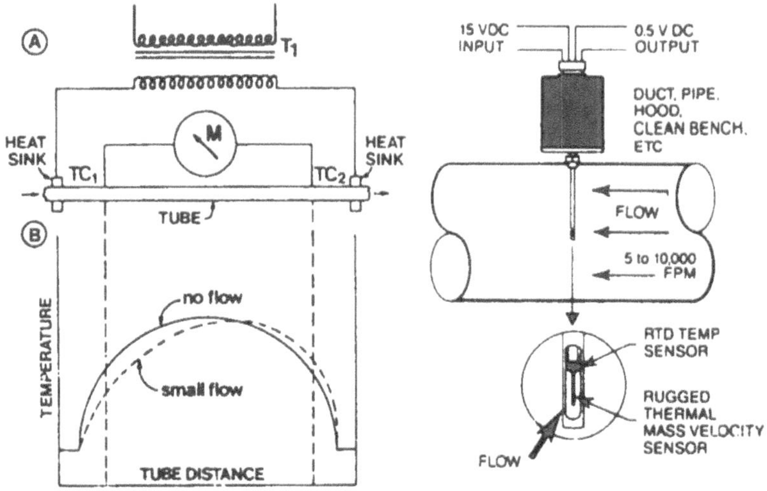

Figure 4.12. Mass flow meters. Reproduced with permission of Omega Engineering, Inc., Stamford, Connecticut.

Mass flow meters are subject to fouling by dirty gases, but are insensitive to changes in pressure and temperature. Mass flow meters should be calibrated against a primary or intermediate standard using the actual gas to be measured.

Inasmuch as mass flow meters measure the rate of mass flow, applications of Boyle's and Charles' Laws to compute standard volume flow rates are never needed. This is the most significant advantage of mass flow meters — the device can be designed to internally convert ideal gas mass flow rate to volume flow rate at standard conditions, using Equation 4.14.

$$Q = \dot{m} \times \frac{R_u \times T_{std}}{P_{std} \times M} = \dot{m} \times \text{ constant} \qquad \textbf{(4.14)}$$

In fact, most mass flow meters read in units of slpm or scfm (standard cubic feet per minute), a great convenience.

Example Problem 4.5

A wet test meter is used to calibrate a mass flow meter using air. The following data are obtained:

V_{wtm} = 30 L
Θ = 9.4 min
P_{wtm} = 745 mm Hg
T_{wtm} = 28 °C
C_{wtm} = 1.002
Mass flow meter reading = MFMR = 3.12 slpm

The two meters are connected in series; thus, with gas volume flow rate corrected to standard conditions and instrument corrections factors incorporated, they should read the same. What is the mass flow meter correction factor, C_{mfm}?

Solution

$$Q_{wtm,\,std} = \frac{V_{wtm} \times C_{wtm}}{\Theta} \times \frac{P_{wtm}}{P_{std}} \times \frac{T_{std}}{T_{wtm}} = 3.10$$

$$C_{mfm} = \frac{Q_{wtm,\,std}}{MFMR} = \frac{3.10}{3.12} = 0.995$$

CHAPTER 4 PROBLEMS

1. Do entire assignment with a spreadsheet; be sure to show equations where appropriate by writing or typing them in as labels or copying to another location and showing this second set of equations as text.
   A rotameter was calibrated for air while the rotameter operated at 730 mm Hg and 21 °C. The following data were obtained:

Rotameter reading (x-axis)	Measured flow rate Q (alpm)
0.5	1.4
1.0	2.0
1.5	2.55
2.0	2.9
2.5	3.55
3.0	4.0
3.5	4.55
4.0	4.75
4.5	5.5
5.0	6.0

   a) Enter the raw data into a spreadsheet.
   b) Correct measured flow rates (Q, alpm) to Q, slpm.
   c) Plot rotameter reading (x-axis) vs Q, slpm.
   d) Find equation of the best-fit line by linear regression.
   e) Use the slope and intercept to compute Q_{best} for each value of x.
   f) Plot Q_{best}.

2. a) The rotameter is later placed in a sampling system where P = 700 mm Hg, T = 50 °C. For each rotameter reading above, compute the Q, slpm at these operating conditions.
   b) Plot these new Q, slpm vs rotameter reading values.
   c) Use linear regression analysis again to find equation of the new line. Write or type equation.
   d) Plot the best-fit line.
3. Now, if when operating at 700 mm Hg and 50 °C, the rotameter reads
   a) 3.0, what is Q, slpm?
   b) 4.2, what is Q, slpm?
   c) 4.8, what is Q, slpm?
4. For each rotameter reading in Problem 3, what is the flow rate in alpm at the rotameter operating conditions?
5. Suppose we wish to use the same rotameter for xenon gas (M = 131.3) instead of air (M = 28.96). Operation is at P = 730 mm Hg and T = 21 °C, and we want 2.5 slpm of xenon. What must the rotameter reading be?
6. A rotameter is calibrated by the manufacturer so that flow rate Q (slpm) at 20 °C and 1 atm is as shown in the following graph.

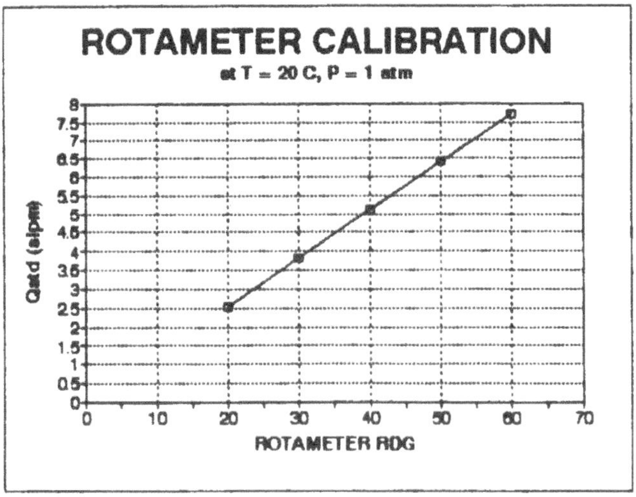

Students calibrate the rotameter at 890 mm Hg and 2 °C and obtain the following data:

RR	Time (min)	V_{wtm} (L)	
20	0.43	1.0	$C_{wtm} = 0.986$
30	0.29	1.0	$T_{wtm} = 21$ °C
40	0.21	1.0	$P_{wtm} = 742$ mm Hg
50	0.17	1.0	
60	0.14	1.0	

a) Plot Q_{std} vs. rotameter reading for these data.
b) If the rotameter is later used at 700 mm Hg, 45 °C, and reads 35, which line should you use — the manufacturer's or the students'?
c) What is the correct value for Q_{std} at this RR of 35?

7. A rotameter is calibrated by the manufacturer so that flow rate Q(acfm) at 20 °C and 1 atm is computed from rotameter reading (RR) by the equation:

$$Q = 0.9 \times RR + 1.07$$

However, the purchaser installs and operates the rotameter at a gage pressure of −4.0 " H_2O, T = 42° C, with P_{bar} = 29.9" Hg.
a) If the rotameter reads 2.0 at the manufacturer's calibration conditions, what is the actual flow rate (acfm)?
b) If the rotameter reads 2.0 at the purchaser's installed condition, what is the actual flow rate (acfm)?
c) What is the installed flow rate in scfm (T_{std} = 25 °C)?

8. The orifice meter equation is Equation 4.5b. EPA, instead of using the coefficient K_m as the unique orifice parameter, specifies ΔH@ as the orifice characteristic, where ΔH@ is the orifice pressure drop that corresponds to a Q of 0.75 scfm (for T_{std} = 68 °F, P_{std} = 29.92" Hg, M_{air} = 29.0). If you know your orifice has K_m = 0.72, what will its ΔH@ be?

9. An orifice was calibrated at 18 °C and 730 mm Hg and a calibration curve for actual conditions was prepared. The calibration data had a best-fit line with the equation:

$$Q_i = 0.244 \times [(\Delta H)^{1/2}] + 0.364$$

where Q is in L/min and ΔH in mm H_2O. The orifice was then used to calibrate a sampling device. In this step, the orifice operated at 25 °C and 720 mm Hg. At the second set of conditions, the orifice ΔH was 63 mm H_2O.

a) Use this ΔH value into the orifice calibration equation to find indicated flow rate Q_i.
b) What was the actual Q_a at sampling device (second set) conditions?
c) What was the standard flow rate?

10. Calculate the flow rate of carbon dioxide through a capillary tube in liters per minute if:

P_u = upstream pressure (absolute) = 18 psi
P_d = downstream pressure (absolute) = 14.7 psi
L = capillary length = 25 mm
T = gas temperature = 20 °C
F = flow constant = 128 (unitless)
υ = gas viscosity: CO_2 = 1.876 × 10^{-4} poise (g/cm-sec)
d = capillary diameter = 0.1 mm

$$Q = \frac{(P_u - P_d) \times \dfrac{[1 + (P_u - P_d)]}{(2 \times P_d)}}{r}$$

where

$$r = (F \times \upsilon \times L) \;/\; (\pi \times d^4)$$

11. The WTM of Problem 3.4 is now used to calibrate a mass flow meter. The mass flow meter reads 4150 sccm (T_{std} = 25 °C) when the WTM is connected in series with it and a pump. The WTM is timed for 3 min and indicates a volume of 13.34 L, while its T = 21 °C and P_{wtm} = −15" H_2O (P_{atm} = 29.5" Hg). Compute C_{mfm}, the mass flow meter calibration factor.

5 VELOCITY MEASURING DEVICES

5.1 INTRODUCTION

A number of flow rate measuring instruments were discussed in Chapter 4. Another way to measure fluid flow rate is to measure its average velocity and multiply by the cross-sectional area of the duct or pipe in which it is flowing. Pitot tubes and electronic velocity meters are often used to measure velocity for this purpose, particularly in air pollutant emission measurement studies. The EPA provides extensive guidance for measuring exhaust duct gas velocities and flow rates. The primary guiding document, found in the Code of Federal Regulations, is 40CFR60, Appendix A, Methods 1 and 2.[2] This information will be covered in detail in Chapter 10.

5.2 PITOT TUBE

The pitot tube is a reliable and adaptable fluid velocity measuring device that has been used for decades. A familiar application is as an airspeed sensor in the wing or nose of an aircraft, but pitot tubes can be used to measure velocity of any flowing fluid. A pitot tube consists of two pressure-sensing tubes connected to a differential pressure gauge. A standard pitot tube is shown in Figure 5.1. The simplicity of the apparatus conceals the complex underlying energy and flow principles involved — a pitot tube can be a precise and accurate velocity sensor.

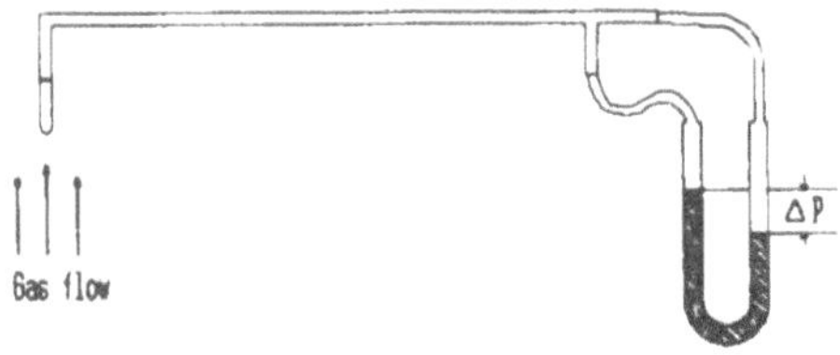

Figure 5.1. Standard pitot tube.

The pitot tube actually measures the difference between the static pressure of the flowing fluid and the pressure due to the velocity or momentum of the fluid molecules. The static pressure, P_{stat}, is defined as the pressure without regard to motion, or the pressure that would be measured if the pressure sensor were moving with the fluid. This static pressure can be measured by inserting an open-tube probe into the fluid so that the opening is parallel to the direction of flow. See Figure 5.2b. In contrast, the pressure due to fluid momentum — called total pressure or static-plus-dynamic pressure — can be measured by inserting an open-tube probe facing directly upstream, as in Figure 5.2a. This pressure is also called "stagnation" pressure because at least some fluid molecules are brought to a complete standstill at the front opening of the probe. If both of the probes are connected to two sides of a differential pressure gauge, as in Figure 5.2c, the gauge reading is the difference in pressures and is called "velocity head":

$$\Delta P = \text{velocity head} = P_{tot} - P_{stat} \tag{5.1}$$

The relationship between the measured velocity head and fluid velocity can be developed from thermodynamic and fluid mechanics principles. To begin, it is assumed that the parcels of fluid that are brought to a halt directly in front of the total pressure port have work done on them adiabatically; that is, the kinetic energy is reduced to zero with no heat flow to surrounding parcels. For this situation, the First Law of Thermodynamics states that the work done must equal the change in all forms of energy of the parcel of fluid.

$$\text{Work} = -(\text{change in all storable energy forms}) + \text{heat} \tag{5.2}$$

The work done as the parcel is halted is related to its specific volume and pressure changes. With the assumption of reversible adiabatic stagnation and for ideal gases, the heat flow is zero, the pressure and specific volume are related through the ratio of specific heats, k:

$$\frac{P_1}{P_2} = \left(\frac{\upsilon_2}{\upsilon_1}\right)^k$$

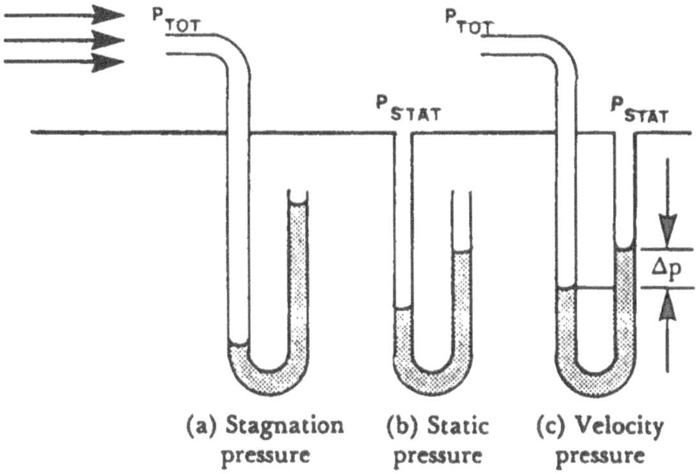

Figure 5.2. Pitot tube operation.

and the work per unit mass can be shown to be a function of the pressures and the density, ρ, which is the inverse of specific volume, υ, of the flowing fluid.

$$w = (P_{tot} - P_{stat})/\rho \quad (5.3)$$

Since the only form of storable energy in Equation 5.2 that changes is kinetic, then,

$$(P_{tot} - P_{stat})/\rho = -(ke) = (v_s^2)/2 - (v_{tot}{}^2)/2 \quad (5.4)$$

where v_s is the velocity of the undisturbed flowing gas, and the kinetic energy per unit mass is $\frac{1}{2} \times v^2$. Also, v_{tot} (velocity at the stagnation point) = 0; thus, the velocity of moving fluid is:

$$v_s = \left[2 \times (P_{tot} - P_{stat})/\rho\right]^{\frac{1}{2}} \text{ or}$$

$$v_s = (2 \times \Delta P_p/\rho)^{\frac{1}{2}} \quad (5.5)$$

where ΔP_p = pitot velocity head = pressure gauge reading.

Using U.S. Customary units, Equation 5.4 must be written as:

$$v_s = \left(2 \times g_c \times \Delta P_p / \rho\right)^{\frac{1}{2}}$$

where g_c = 32.2 lbm-ft/sec²/lbf to make the units cancel. When the pitot tube is used to measure velocity of a gas, the Ideal Gas Law may be employed to express density:

$$\rho = P_{stat} \times M_{gas} / R_u \times T_{gas} \tag{5.6}$$

thus

$$v_s = C_p\left\{\left(2 \times g_c \times R_u \times T_{gas} \times \Delta P_p\right) / \left(P_{stat} \times M_{gas}\right)\right\}^{\frac{1}{2}} \tag{5.7a}$$

or

$$v_s = K_p \times C_p \times \left(\frac{T_{gas}}{P_{stat} \times M_{gas}} \times \Delta P_p\right)^{\frac{1}{2}} \tag{5.7b}$$

where $K_p = (2 \times g_c \times R_u)^{1/2}$ + units corrections and

C_p = a coefficient to account for turbulence and wake effects, if any.

A common set of units is: inch H_2O for ΔP_p
inch Hg for P_{stat}
°R for T_{gas}
ft/sec for v_s
lb/lbmol for M_{gas}
For these units, K_p = 85.49

Using metric units, one obtains: mm H_2O for ΔP_p
mm Hg for P_{stat}
K for T_{gas}
m/sec for v_s
g/gmol for M_{gas}
For these units, K_p = 34.97

A "standard" pitot tube is constructed with the two pressure-sensing tubes concentrically mounted as shown in Figure 5.3. The inner tube measures total pressure, while the static pressure is measured by numerous small holes in the side of the outer tube. These small holes are located a distance six times the outer diameter back from the tip, where there is very little turbulence.

If a standard pitot tube is constructed in accordance with Figure 5.3, the flow pattern is such that C_p (the coefficient in Equation 5.7) is 0.99 to 1.0, with 0.99

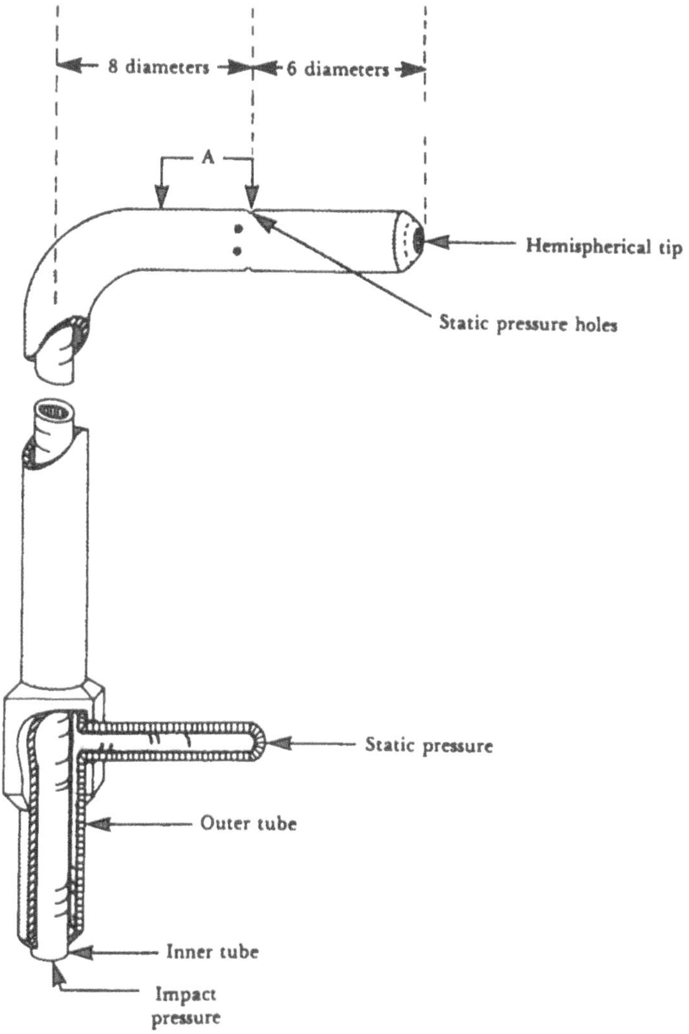

Figure 5.3. Standard pitot tube (Reference 15).

the commonly used value. Standard pitot tubes are frequently constructed of stainless steel with outside diameters from about 3/32 to 5/16" and any convenient length of the long leg from a few inches to 6 or 8 ft. Standard pitot tubes have been used extensively in ventilation duct air velocity measurement and flow balancing though, in recent years, insertion style mass flow meters have become more commonly used for these purposes.

One drawback of standard pitot tubes is their propensity to plug up in gas streams with high dust loading. In addition, the amount of velocity head produced with a standard pitot is small — about 0.05" H_2O for 1000 ft/min (5 m/sec). For accurate readings, an inclined water manometer or a Hook gauge

Figure 5.4. Inclined manometer and hook gauge.

(Figure 5.4) is required. In low-velocity situations, the standard pitot may not produce enough velocity head to be readable.

A modification to the standard pitot design was developed to overcome the problem of plugging in dirty gas streams and to slightly enhance the relationship between velocity and ΔP_p — the Stausscheibe or S-type pitot.

S-type pitot tubes are designed as in Figure 5.5, generally of $^3/_{16}$ to $^3/_8$" stainless steel tubing. The pressure-sensing openings are large enough to avoid plugging. In addition, the S-type pitot is easier to use than a standard pitot for three reasons: 1) it can be more rugged; 2) it is essentially straight, and therefore easier to insert into a duct or pipe; and 3) since one port faces directly downstream, there is a slight aerodynamic low-pressure wake at the opening, resulting in a pressure that is lower than static pressure and a ΔP_p larger and more readable than that produced by a standard pitot.

Equation 5.7 holds for an S-type pitot tube, where the coefficient $C_{p(s)}$ accounts for the above-mentioned wake effects and must be determined by calibration against a standard pitot. The calibration involves measuring the same gas stream with both styles of pitot. See Figure 5.6.

From Eqution 5.7b, the velocity for each pitot tube is the same; that is:

$$v_s = C_{p(std)} \times K_p \times \left\{\left(T_{gas} \times \Delta P_p\right) / \left(P_{stat} \times M_{gas}\right)\right\}^{\frac{1}{2}}$$

$$= C_{p(s)} \times K_p \times \left\{\left(T_{gas} \times \Delta P_p\right) / \left(P_{stat} \times M_{gas}\right)\right\}^{\frac{1}{2}} \quad (5.8)$$

Solving for $C_{p(s)}$ (the coefficient for the S-type pitot), one obtains:

$$C_{p(s)} = C_{p(std)} \times \left(\Delta P_{std} / \Delta P_s\right)^{\frac{1}{2}} = 0.99\left(\Delta P_{std} / \Delta P_s\right)^{\frac{1}{2}} \quad (5.9)$$

Example Problem 5.1. S-type Pitot calibration calculations

An S-type pitot is calibrated as in Figure 5.6. Three different gas velocities are employed, and the following data obtained:

ΔP_{std}	$\Delta P_{s\text{-type}}$(leg A)	$\Delta P_{s\text{-type,reversed}}$(leg B)
0.66	0.88	0.90
0.72	0.99	1.04
0.84	1.1	1.17

All values in inches of H_2O.

Compute the S-type pitot coefficient for each orientation.

Solution

$C_{p(s)} = 0.99 \times (0.66/0.88)^{1/2} = 0.857$
$C_{p(s)} = 0.99 \times (0.72/0.99)^{1/2} = 0.844$
$C_{p(s)} = 0.99 \times (0.84/1.17)^{1/2} = 0.839$
Avg C_p: 0.847

Reversed $C_p(s) = 0.99 \times (0.66/0.90)^{1/2} = 0.848$
Reversed $C_p(s) = 0.99 \times (0.72/1.04)^{1/2} = 0.824$
Reversed $C_p(s) = 0.99 \times (0.84/1.19)^{1/2} = 0.832$
Avg C_p: 0.835

Each S-type pitot must be individually calibrated, and recalibrated regularly to check for changes due to tube distortion. EPA Method 2[2] is covered in Chapter 10 and gives specific instructions for calibration equipment and set-up. An S-type pitot constructed as in Figure 5.5 should have a value for $C_{p(s)}$ of approximately 0.84, with either leg facing upstream. The exact value of the coefficient will depend on pitot shape, dimensions, and the extent to which the faces of the openings are parallel.

Example Problem 5.2.

Compute velocity and flow rate from pitot data. Using the pitot of the previous Example Problem in the reversed position, velocity measurements

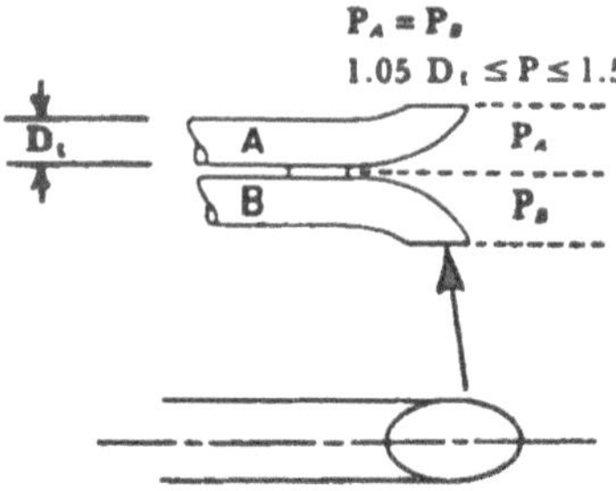

Figure 5.5. S-type pitot details.

were made in a 3 × 3-ft square duct. Four equally spaced points in the cross-sectional area of the duct were measured, and the following pitot pressure data were obtained.

1. 0.70" H_2O
2. 0.75
3. 0.72
4. 0.77

The gas in the duct had T = 200 °F, P = 29.5" Hg static pressure, and molecular weight = 27.7.

Solution

From Equation 5.7b:

$$-.835 \times 85.49\left\{\left[(200 + 460) \times \Delta P_p\right]/(29.5 \times 27.7)\right\}^{\frac{1}{2}}$$

$$= 64.154 \times \left\{\Delta P_p\right\}^{\frac{1}{2}}$$

or

the four velocities are:

1. 53.7 ft/sec
2. 55.6 ft/sec
3. 54.4 ft/sec
4. 56.3 ft/sec

or Avg Velocity: 55.0 ft/sec

The duct flow rate is the velocity times the cross sectional area or 495 ft^3/sec = 29,700 actual ft^3/min (acfm).

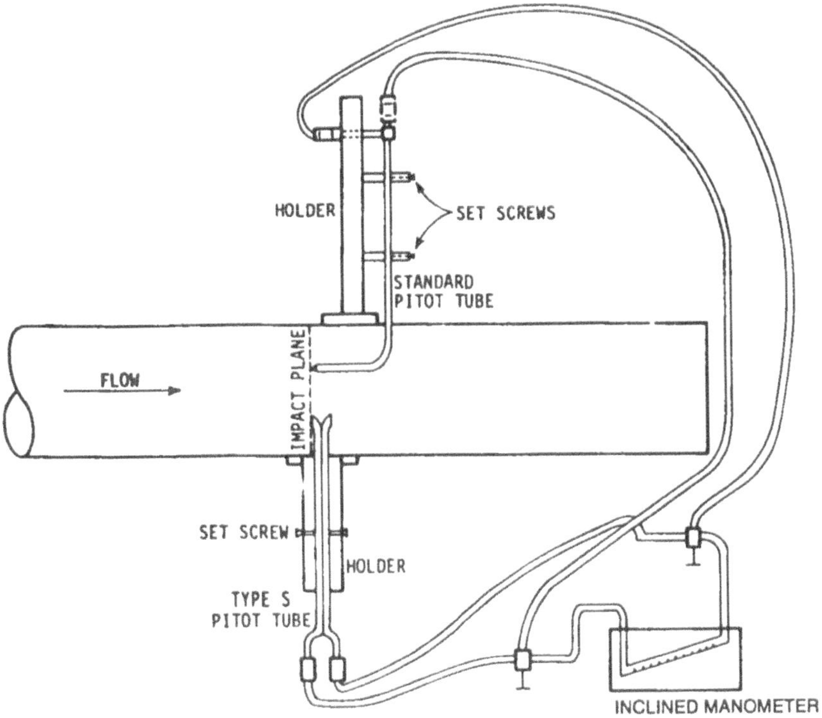

Figure 5.6. S-type pitot calibration (Reference 12).

5.3 ELECTRONIC VELOCITY METER

Insertion-style mass flow meters, as discussed in Chapter 4, are packaged by several manufacturers as electronic velocity meters. As shown in Figure 5.7, the sensor portion of the flow meter is a slender probe that can be inserted into the duct to measure two temperatures. One thermocouple or RTD temperature sensor measures the actual gas temperature, and the other is heated to a specific temperature above the first. The flowing gas cools the second sensor at a rate related to the heat transfer properties of the probe and the gas. Energy supplied to the heated sensor to maintain its temperature above that of the gas is proportional to the cooling rate, which is in turn proportional to the mass flow rate of the gas.

As noted in Chapter 4, from the mass flow rate of an ideal gas, the volume flow rate at standard temperature and pressure can be easily computed. The gas temperature sensor built into the probe allows an approximate correction to actual conditions in the duct. The gas flow rate is the rate of gas flowing through the "window" in the probe (see Figure 5.7); and since this area is fixed, the velocity is found by dividing flow rate by area.

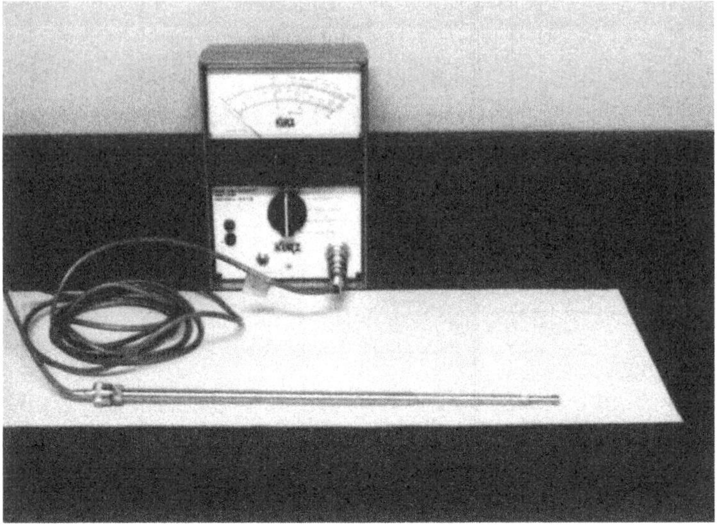

Figure 5.7. Electronic velocity meter.

All of the calculations can be done electronically, so the meter can be read directly in velocity units. Several manufacturers produce portable, hand-held models with rechargeable batteries — very convenient devices for duct velocity measurements. Arrays of sensors can be "ganged together" on a single probe for simultaneous measurement of velocity at a number of positions across the duct.

As noted in Chapter 4, the biggest drawback of electronic velocity meters is their sensitivity to damage or error in dirty or corrosive gas streams.

CHAPTER 5 PROBLEMS

1. An S-type pitot tube was calibrated with both leg A upstream, then leg B upstream against a standard pitot which has a coefficient of 0.99. For four different velocities, the following pressure data were obtained:

Velocity	Std. Pitot ΔP (" H_2O)	S-type leg A ΔP_a (" H_2O)	S-type leg B ΔP_b (" H_2O)
1	0.5	0.68	0.67
2	0.42	0.56	0.57
3	0.36	0.49	0.46
4	0.20	0.26	0.27

a) Compute both S-type pitot coefficients for each velocity, $C_{p(a)}$ and $C_{p(b)}$.

b) Compute average $C_{p(a)}$ and $C_{p(b)}$ values and the difference.

2. For a certain stack, the exhaust gas molecular weight = 30.02, the pitot coefficient = 0.84, the gage static pressure in the duct = -1.3" H_2O, the duct diameter = 3 ft, the average stack temperature = 630 °R, P_{atm} = 29.92" Hg, and moisture = 0.
There are 12 sampling points (6 on each of two perpendicular diameters).
The measured ΔP values were:

Horiz. diameter (" H_2O)	Vert. diameter (" H_2O)
1. 0.30	0.28
2. 0.35	0.40
3. 0.65	0.76
4. 0.80	0.72
5. 0.20	0.25
6. 0.20	0.20

Find average velocity and compute Q_s, the volume flow rate in standard cubic meters per hour (scmh). Think about units!

3. Given the pitot velocity equation:

$$v_s = K_p \times C_p \times \left[T_{gas} / \left(P_{stat} \times M_{gas}\right) \times \Delta P_p\right]^{\frac{1}{2}}$$

where K_p is a constant whose value depends on the units of the various parameters, find K_p for the following:
a) T_{gas} in K
b) P_s in kPa
c) M_{gas} in g/gmole
d) ΔP in inches H_2O
e) v_s, stack gas velocity, in m/sec
The easy way, but not the only way, is to start with the known value of K_p = 34.97 m/sec × [(g/gmole × mm Hg)/(K × mm H_2O)]$^{1/2}$

4. A pitot tube is calibrated in the wind tunnel.
T_{atm} = 24 °C, P_{atm} = 29.7" Hg. The following three sets of data are obtained: (ΔP units are inches H_2O):

Run #	ΔP_{std}	ΔP_s-type Leg A	Ps, gage (" H_2O)	T_{gas} (°C)
1	0.66	0.80	−1.7	24
2	0.77	0.92	−1.9	24
3	0.84	1.07	−2.3	24

a) Compute the average pitot coefficient for this leg.
b) Compute the average deviation for this leg.

6 FLOW MOVING AND CONTROLLING DEVICES

6.1 INTRODUCTION

In many applications in air sampling, it is important to maintain a constant flow rate for a period of time. The devices discussed in Chapters 4 and 5 allow the instantaneous measurement of flow rate, and an operator can make adjustments to the pump speed or change the inlet resistance to the air mover by adjusting a valve to maintain a steady flow rate based on observation of the measurement device's output. The operation of air movers will be discussed in Section 6.5.

However, it is often more convenient and efficient to utilize automatic flow-controlling equipment. In this chapter, several types of flow control devices will be discussed: critical orifices, electronic velocity or mass flow sensors, and differential pressure flow controllers. Current applications of all three types to air sampling are found in PM10 samplers, personal pumps used for indoor air studies, and a large variety of ambient air monitoring and sampling instruments.

6.2 CRITICAL OR LIMITING ORIFICE

As mentioned in the discussion on orifice meters in Section 4.3, the relationship of orifice meter flow rate to pressure drop is valid only when there is a

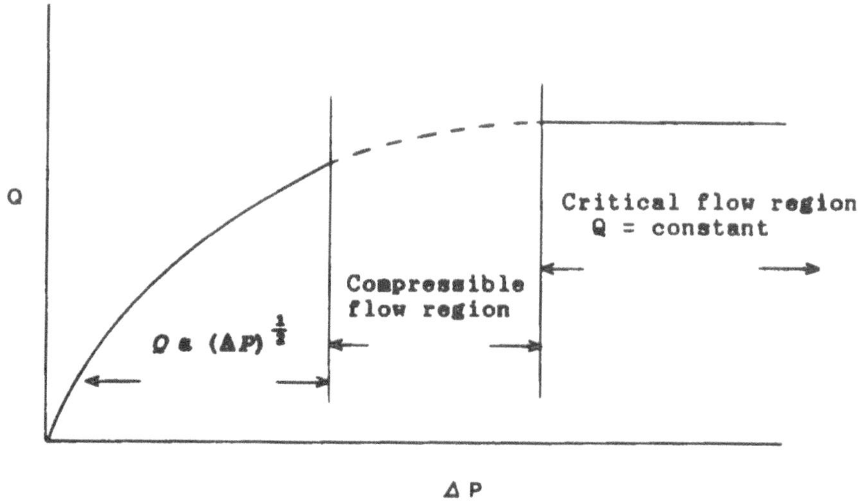

Figure 6.1. Critical orifice flow rate curve.

small enough pressure drop that the fluid acts incompressibly. If the pressure drop across an orifice is relatively large, the change in gas density cannot be ignored and Equation 4.5 and Figure 4.4 are not valid. However, if the pressure drop is very large (and especially for fairly small orifices — orifice area less than $^1/_{25}$ of the upstream area[46]), the velocity in the throat of the orifice reaches sonic velocity. When this occurs, the flow rate reaches its maximum, and further decreases in downstream pressure will not affect the flow rate.[45] Therefore, small variations in pump vacuum do not cause the flow rate to change and the orifice is a "constant flow controller". Figure 4.4a can be expanded to include this sonic or "critical" region as shown in Figure 6.1.

The following questions surely arise: what is the minimum pressure drop across the orifice necessary to achieve sonic velocity or critical flow in the orifice? How great is the critical flow? Another question that follows immediately is: how does one measure flow?

The answer to the first question has been developed empirically, and researchers disagree somewhat. Perry (Chemical Engineers Handbook)[46] and Marks (Mechanical Engineers Handbook)[47] state that sonic velocity for air is reached when the downstream absolute pressure is less than 0.53 times the upstream pressure.

$$P_{downstream} \leq 0.53 \times P_{upstream} \tag{6.1}$$

Corn and Bell[44] and Lodge et al.[43] found, using air and hypodermic needles as orifices, that downstream pressure must be less than 0.45 times upstream pressure or

$$P_{downstream} \leq 0.45 \times P_{upstream} \qquad \textbf{(6.2a)}$$

or pressure drop across the orifice:

$$P_{up} - P_{down} = \Delta P_{orifice} \geq 0.55 \times P_{up} \qquad \textbf{(6.2b)}$$

Equation 6.2a or 6.2b is a bit more conservative a criterion than Equation 6.1.

Each orifice has a unique and nonadjustable critical flow rate. The air sampling personnel must select a large-enough diameter orifice to provide the approximate flow rate desired. In practice, each critical orifice must be flow-checked in the critical region. In performing this single-point calibration, and in using the critical orifice to control flow rate, one can ensure critical flow by ensuring that the system meets the more restrictive of the above two criteria: i.e., Equation 6.2.

If upstream pressure is approximately atmospheric, downstream absolute pressure must be less than about 0.45 × 30" Hg or 13.5" Hg (243 mm Hg). Since a mercury manometer or vacuum gauge are the most frequently used devices to monitor downstream pressure, it is convenient to report the vacuum reading. The downstream vacuum (gage pressure) should be more than about 17" Hg (417 mm Hg) below atmospheric pressure to ensure critical flow. Sampling train components upstream of the orifice may decrease the upstream pressure below atmospheric; thus, Equation 6.2 should be employed to check the minimum critical pressure. The simplest expedient is to ensure that the vacuum reading on the downstream side of the orifice is a few inches of mercury greater than the minimum.

If the minimum downstream vacuum gauge reading for critical flow is "Vac", Equation 6.2a can be written:

$$P_{down,abs} = P_{atm} - Vac \leq 0.45 \times P_{up}$$

or

$$Vac \geq P_{atm} \times 0.45 \times P_{up}$$

and if $P_{up} = P_{atm}$ – (sampling train head loss, ΔH_s), then

$$Vac \geq 0.55 \times P_{atm} + 0.45 \times \Delta H_s$$

For example, if $P_{atm} = 745$ mm Hg and sampler components cause a head loss of $\Delta H_s = 35$ mm Hg, the downstream vacuum must be at least 0.55 × 745 + 0.45 × 35 = 426 mm Hg. Thus, a downstream vacuum of about 475 mm Hg or around 19" Hg would provide a margin to safely ensure critical flow.

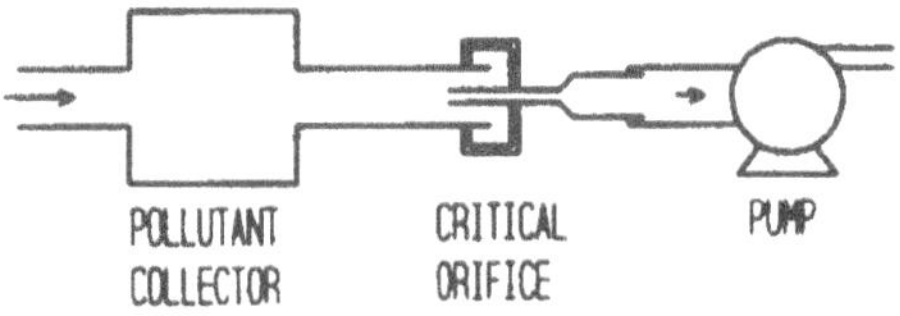

Figure 6.2. Typical critical orifice setup.

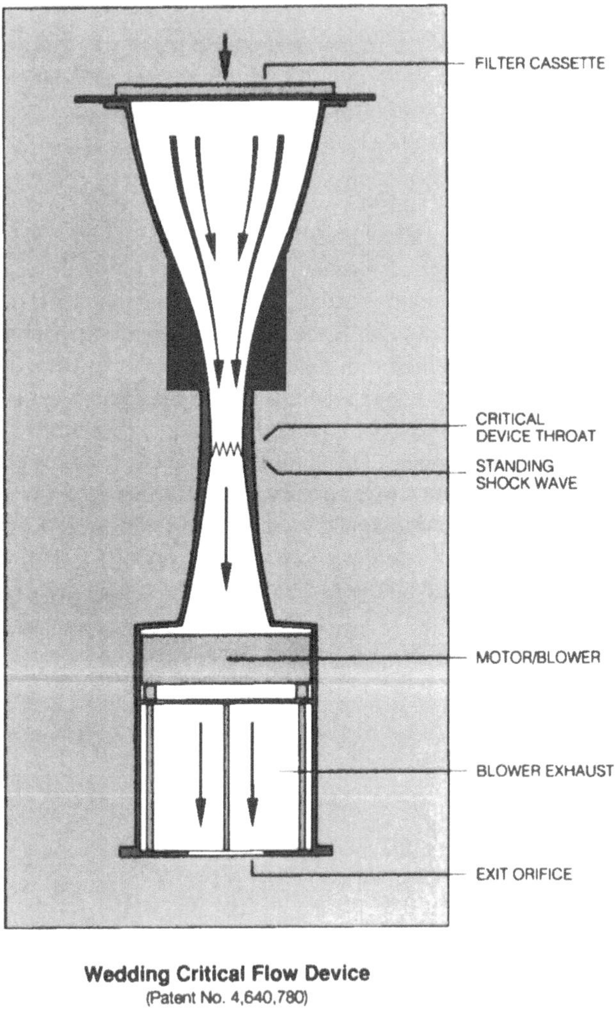

Figure 6.3. Wedding PM10 critical flow control. (Courtesy Wedding & Associates.)

Now, to return to the third question posed above: how does one evaluate critical flow rate? Usually, a soap-bubble meter is employed at the inlet or exit of the sampling train containing the critical orifice. The orifice is then operating at the same conditions during calibration as it will experience during sampling *(in situ)*. With the downstream vacuum set above the minimum required, the critical orifice standard flow rate is calculated from Equation 3.5.

$$V_{std} = V_{meas} \times \frac{(P_{bar} - P_{vap})}{P_{std}} \times \frac{T_{std}}{T_{meas}} \tag{3.5}$$

Critical orifices, like any orifice, are affected by changes in pressure, temperature, or molecular weight of the gas passing through the orifice. Therefore, if the orifice pressure or temperature during use differs from the calibration conditions, the actual flow rate must be calculated using Equation 4.6, repeated here as Equation 6.3 for the assumption of no change in molecular weight of the gas measured:

$$Q_{std,\,field} = Q_{std,\,cal} \times \left(\frac{P_{cal}}{P_{field}} \times \frac{T_{field}}{T_{cal}} \right)^{\frac{1}{2}} \tag{6.3}$$

It is worth emphasizing again that the square root function evolves from orifice theory and is applied any time an orifice operates at conditions other than those at calibration. Equation 6.3 should not be confused with an application of Boyle's and Charles' Laws, as in Equation 3.5.

Because critical orifices are small, they must be protected from plugging with particles in the gas stream. Critical orifice accuracy is on the order of ± 2%. Hypodermic needles have long been employed in air sampling trains; — they are easy to use, inexpensive, and fairly uniform in size and flow rate. Twenty-seven gauge needles give a critical flow rate around 200 mL/min and were the standard for bubblers used in the 1970s to monitor ambient sulfur dioxide.[2] More recently, microliter syringes have been employed as critical orifices for much smaller flow rates in indoor air pollution studies.

Figure 6.2 shows a typical arrangement of a critical orifice in a low-flow air sampling system. Figure 6.3 is a much higher flow application in a PM10 ambient particulate sampler.

Example Problem 6.1

A critical orifice was calibrated with air at 21 °C and 750 mm Hg. It is to be used in an air sampling system with an inlet temperature of 24 °C, pressure

745 mm Hg, sampling train head loss of 45 mm Hg, and a gas temperature just upstream of the orifice of 24 °C. The measured calibration flow rate is reported as 0.85 sccm.

a) What is the flow rate at the use conditions in sccm?
b) What is the flow rate at use conditions in accm?

Solution

a) The calibrated flow rate is given corrected to standard conditions (25 °C, 1 atm); thus, Equation 6.3 is employed:

$$Q_{std,\ field} = 0.85 \times \left[750/(745 - 45) \times (297/294)\right]^{\frac{1}{2}} = 0.88\ \text{sccm}$$

b) Assuming the flow rate of interest is the actual rate at 24 °C and 745 mm Hg entering the sampling train inlet, Boyle's and Charles' Laws are used to convert the answer from part a), which is at 25 °C, and 760 mm Hg, to these conditions:

$$Q_{act} = Q_{std} \times (760/745) \times (297/298) = 0.895\ \text{accm}$$

6.3 ELECTRONIC FLOW CONTROLLERS

Electronic flow controllers are basically an application of electronic velocity meters discussed in Chapter 5 and insertion-type mass flow meters discussed in Chapter 4. A hot-wire, constant-temperature anemometer or mass flow sensor is inserted into the gas stream. As described in Section 5.3, gases passing across the heated element have a cooling effect proportional to the mass rate and the type of gas flowing past the sensor. The gas is assumed to be ideal, and the mass of gas passing through the sensor is a product of the density and velocity of the gas.

The control system of the flow controller is designed for a particular molecular weight (usually, that of air). A separate gas temperature measurement allows the control system electronics to maintain the heated element at a constant temperature above that of the gas. Since the heat transfer rate from the hot probe to the gas is directly proportional to the temperature difference and to the velocity of the gas, maintaining this constant difference is essential. A change in temperature of the hot sensor thus indicates a change in velocity of the gas as might happen, for example, if there were an increase in head loss in the portions of the sampling train upstream of the sensor.

The probe and related electronic circuitry produce an output signal that is proportional to the change in gas velocity. The feedback control circuit contains

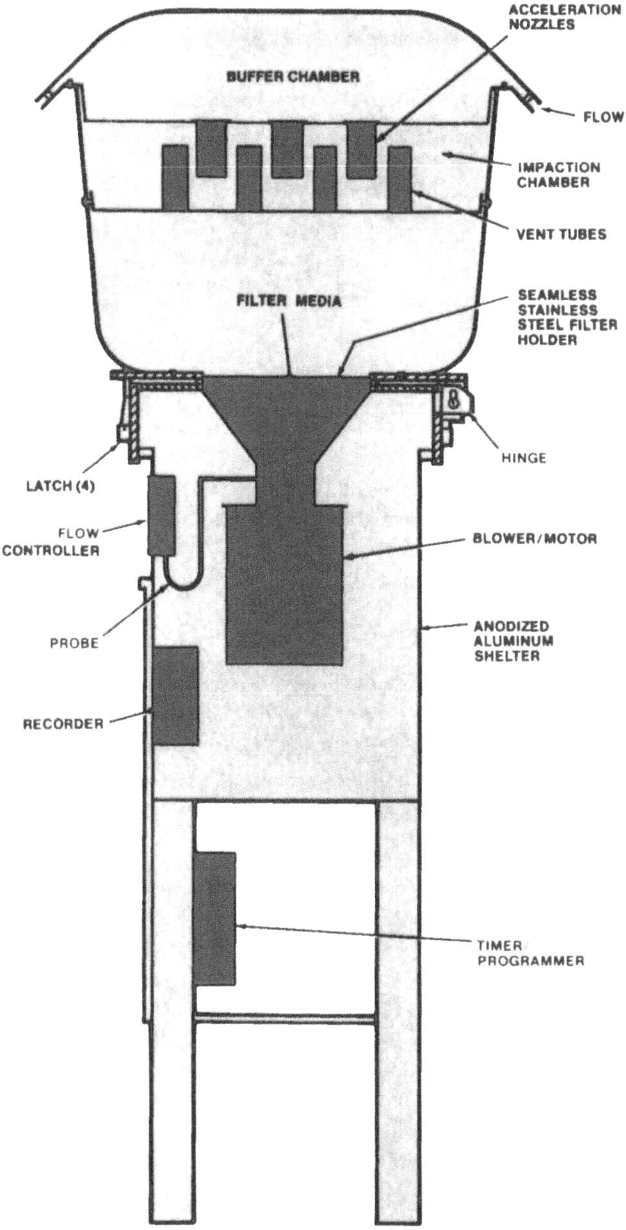

Figure 6.4. Electronic flow controller in PM10 sampler. (Courtesy of Graseby-Andersen, Atlanta, Georgia.)

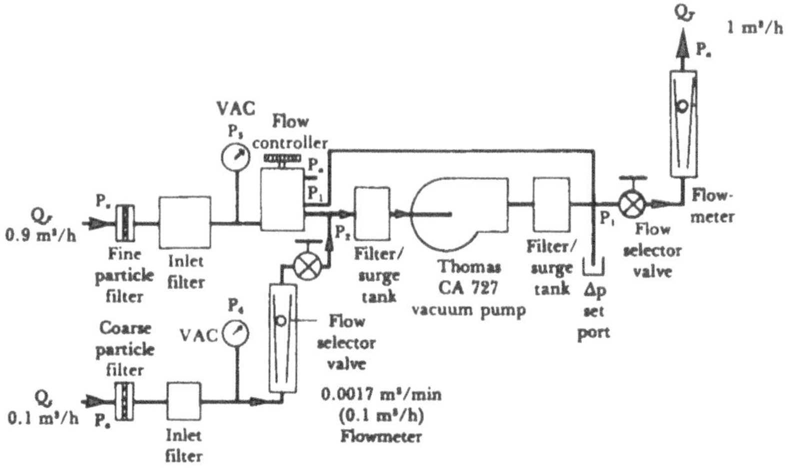

Figure 6.5. Differential pressure flow controller. (Courtesy of Graseby-Andersen, Atlanta, Georgia.)

the electronics necessary to vary the voltage to a pump or air mover so as to adjust the flow rate up or down in order to maintain a constant, preselected flow rate.

Many total suspended particulate ambient samplers and PM10 Samplers utilize this type of flow control to maintain a steady flow rate as particulate matter builds up on the filter over the twenty-four hour sampling period. See Figure 6.4. Electronic flow controllers maintain their calibrated settings well, but must be occasionally checked for dust build-up on the sensor that would change heat transfer characteristics.

6.4 DIFFERENTIAL PRESSURE FLOW CONTROLLERS

Differential pressure flow controllers allow the desired flow rate to be set by the operator by means of an adjustable valve in the flow path, usually downstream of the sample collector. Partially closing the valve causes a pressure drop, which is sensed by a pressure transducer. Should the pressure drop decrease, indicating lower flow, the pressure sensor operates a controller either electronically or mechanically to increase flow. Depending on the application, this is accomplished by increasing the pump speed or opening a throttle valve in the flow path. Many modern portable pumps use differential pressure to control DC motor speed and hence flow rate. In either case, the flow rate is increased until the original pressure drop is reestablished. Figure 6.5 shows a flow diagram of a typical differential pressure flow control system utilizing a flow path throttle.

6.5 AIR MOVERS AND PUMPS

The air mover is an essential part of any active air sampling system — only diffusion samplers as used for radon, formaldehyde, and some personnel monitors used for exposure evaluation can function without an air mover. The air mover must cause the air or gas to move into and through the sample collection medium, where the contaminant of interest is removed. Depending on the application, the air mover must cause this air flow for a specified volume, rate, or time. The selection of an air mover depends upon the characteristics of the contaminant being sampled, the rate or quantity of air to be sampled, the resistance to flow to be overcome, the availability of power, the need for portability, noise, reliability, and maintenance considerations, and a number of other considerations.

A pump to be used to check for occupational exposure to air pollutants may have to sample at 20 mL/min for 8 hr while being carried by an employee throughout a routine workday. Obviously, primary considerations for such a pump are weight, battery life, noise, and controlled sampling rate. A PM10 sampling pump may operate for 24 hr at a flow rate of 1.13 m^3/min and be required to do this every sixth day for a year. Such a pump would need to be reliable, rugged, and have a high flow rate; weight and noise would be secondary considerations. Ambient sampling for volatile organic compounds using EPA's TO-14 protocol[11] requires the use of an evacuated 6 L canister. The canister's initial vacuum is the air-moving force, and a specified volume of air is sampled. EPA Method 5 emission monitoring[2] requires a pump that can maintain 15" Hg vacuum and sample for several hours at a flow rate of 0.021 m^3/min.

Clearly, a thorough understanding of the sampling objectives, regulatory guidelines, and the head losses and required flow rates of the sampling apparatus is prerequisite to the selection of an air mover. The basic categories from which to choose are:

1. Positive displacement pumps (piston and diaphragm)
2. Centrifugal pumps
3. Evacuated canisters
4. Liquid displacement flasks

Positive Displacement Pumps

In a piston pump, gas is drawn into a cylinder on the suction stroke of a piston. An intake valve is operated by a cam to open during this stroke and allow gas to be drawn in. At the bottom of the stroke, the intake valve closes and an exhaust valve opens, as shown in Figure 6.6. A battery-operated or AC motor, or an internal combustion engine, can power the pump. Piston pumps range in size from small, portable, personal pumps with flows from 2 to 5000

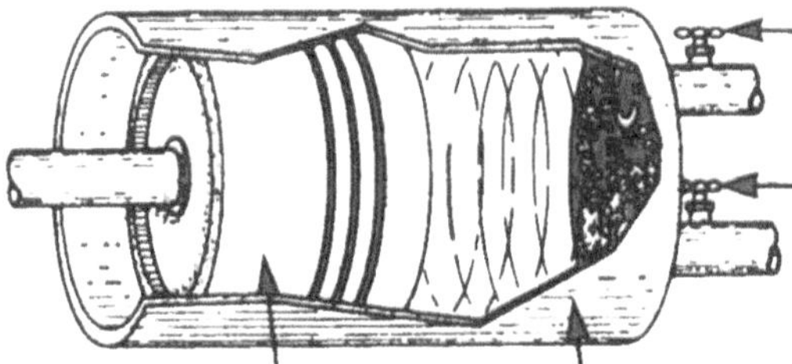

Figure 6.6. Piston pump (Reference 15).

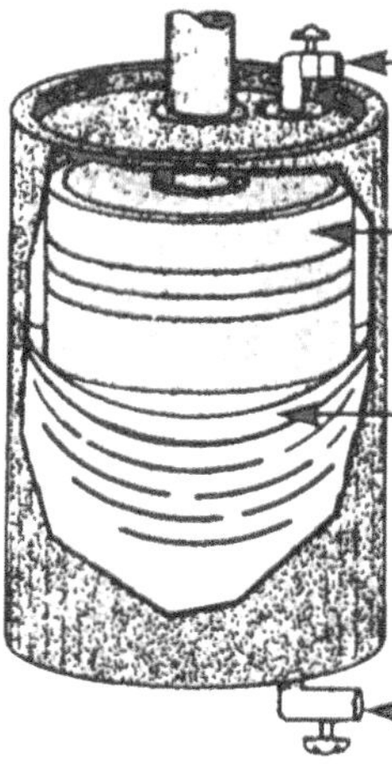

Figure 6.7. Diaphragm pump (Reference 15).

mL/min, to very large air compressors having flow rates of many cubic meters per minute. Well designed piston pumps are the only acceptable choice for drawing a container down to a very high vacuum.

Diaphragm pumps operate in a fashion similar to that of piston pumps, but are generally fairly low-flow rate devices. A flexible diaphragm is attached to a movable plunger or eccentric crankshaft. The diaphragm flexes as the motor turns the crankshaft, creating alternating pressure and vacuum. Valves are typically neoprene checkvalves in air sampling pumps, and operate with the pressure changes to allow alternating intake and exhaust of gas. See Figure 6.7. Diaphragm pumps can be designed with very few and very light-weight parts, and thus are often the choice for portable pumps. Diaphragm pumps cannot produce high flow rates or high vacuums.

Positive displacement pumps of both varieties usually have a negative linear relationship between head produced (the difference between inlet and outlet

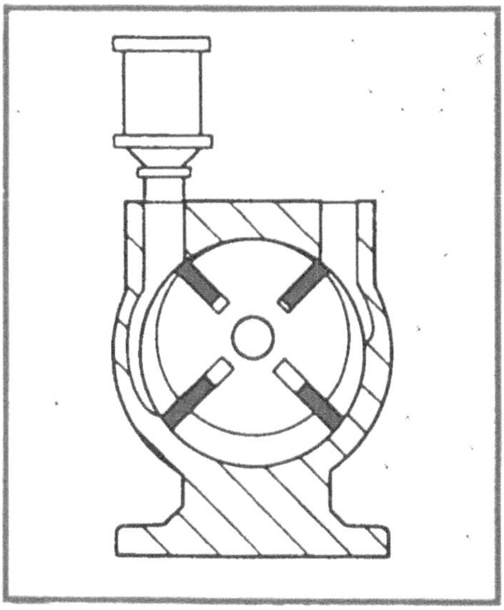

Figure 6.8. Sliding vane rotary pump.

absolute pressure) and pump capacity; as the flow resistance or head loss at the inlet increases, and absolute inlet pressure decreases, the pump's flow rate decreases linearly.

Centrifugal Pumps

Centrifugal pumps include common rotary vane pumps as well as fans and blowers. They are probably the most common type of pump for handling liquids and gases. For air sampling, the most frequently used models consist of an impeller with sliding carbon or phenolic and fiber vanes. The vanes are pushed against the outer casing by centrifugal action and wear in to provide a tight seal. The rotation of the impeller creates a low pressure at the air intake port. Air drawn into the vane area is accelerated to high velocity by the rotation of the impeller (typically at 1700 to 3500 rpm), and then discharged at higher pressure out the outlet. This type of centrifugal pump is commonly called *radial flow design* and is the type used in most air pollutant emission monitoring equipment (see Figure 6.8).

Fans and blowers used in high volume and PM10 samplers are axial flow centrifugal pumps, where rotating vanes or blades primarily cause air to be drawn and pushed along the axis of the pump's rotor.

Centrifugal pumps do not have a linear relationship between head produced and capacity or flow rate. A pump must be selected that has the required capacity at the pressure head that will be encountered. In most air sampling applications,

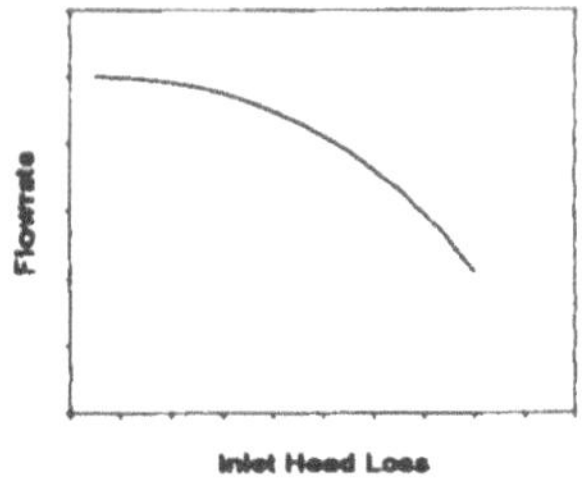

Figure 6.9. Centrifugal pump capacity.

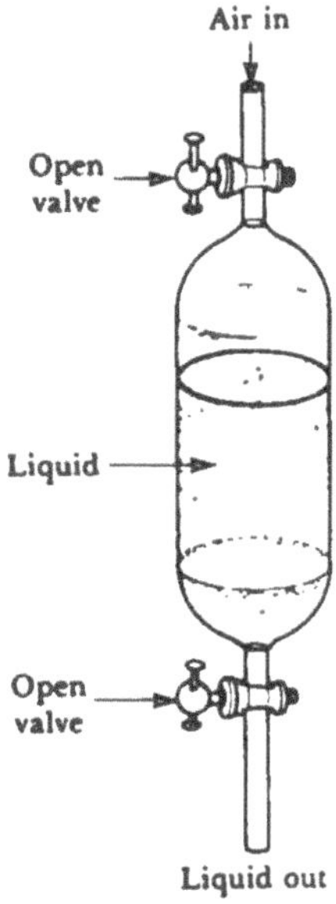

Figure 6.10. Liquid displacement flask.

the pump operates at a fixed outlet pressure; if inlet head loss increases during the sampling process, the capacity or air flow rate decreases in a nonlinear fashion, as in Figure 6.9. If constant flow rate is desired, some means of flow control, as discussed earlier in this chapter, must be employed.

Evacuated Canisters

Modern air toxics sampling methods (e.g., EPA's TO-14)[11] specify the use of polished stainless steel canisters evacuated to absolute pressure of less than 0.05 mm Hg as air movers. With a flow-restrictive inlet such as a mass flow controller or critical orifice, the evacuated canister can sample at slow rates, without external power, for sampling periods as long as 24 hr.

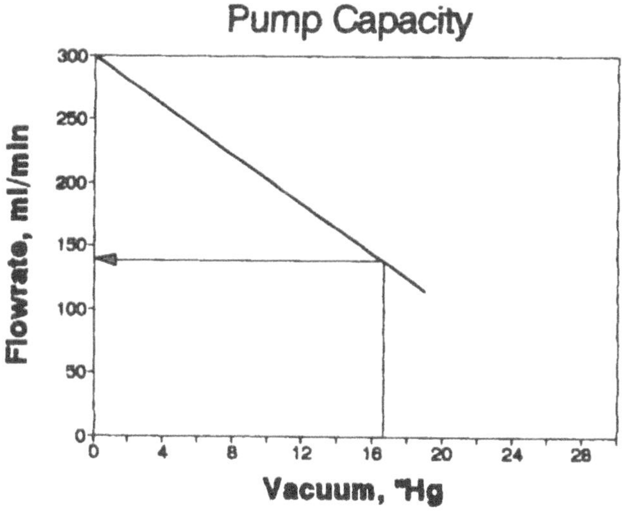

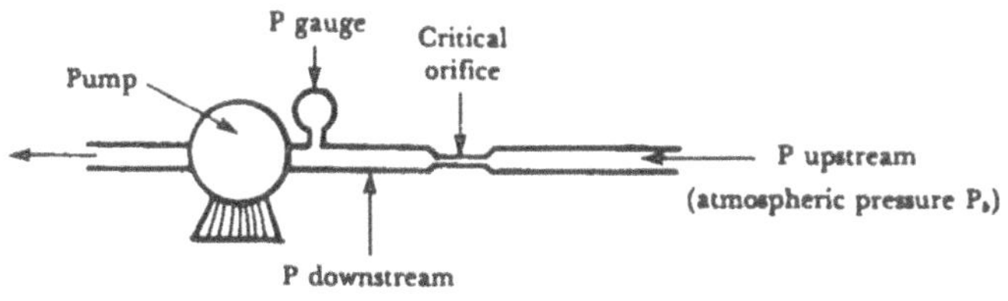

Figure 6.11. Piston pump capacity.

Liquid Displacement Flasks

Liquid displacement flasks operate on the same principle as the displacement bottle of Section 3.4 in Chapter 3. Water or other liquid is drained from a flask of known volume; an equal volume of air is drawn into the flask. See Figure 6.10.

Example Problem 6.2

A piston pump has capacity as shown in Figure 6.11. An air sampling specialist wishes to use this pump to sample with a critical orifice flow controlling device. If the pressure upstream of the orifice is atmospheric or 29.9" Hg, can the pump be used for:

a) A critical orifice at 50 mL/min?
b) A critical orifice at 200 mL/min?

Solution

The requirement for critical flow is given by Equation 6.2 as:

$$P_{downstream} \leq 0.45 \times P_{upstream}$$

Here,

$$P_{downstream} \leq 0.45 \times 29.9 = 13.5'' \text{ Hg}$$

$$P_{gage} = 13.5 - 29.9 = -16.4 \text{ or } 16.4'' \text{ Hg vacuum}$$

The plot shows that at a vacuum of 16.4" Hg, the pump capacity is only 140 mL/min.

CHAPTER 6 PROBLEMS

1. A manufacturer has developed a PM10 sampler that uses a "critical device throat" between the sample filter and the motor for flow control, as shown in the sketch below. One experimental blower motor they are trying out maintains a vacuum of 17.4" Hg at the blower inlet when P_{atm} = 28.6" Hg and ΔP_{filter} = 2.5" Hg.

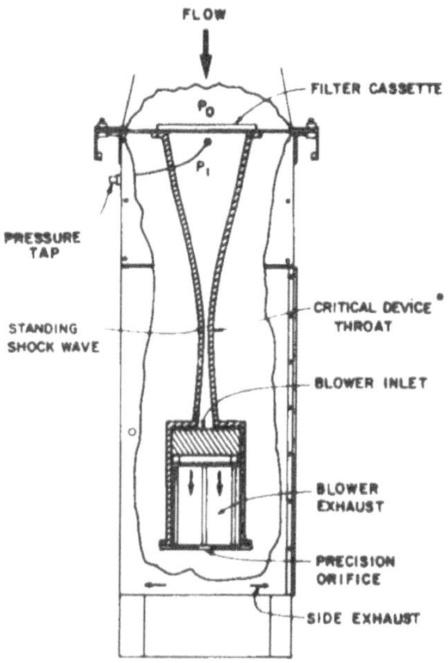

Courtesy of Wedding & Associates, Ft. Collins, Colorado.

According to limiting orifice theory, is the flow critical under these conditions?

2. A gas sampler that uses a critical orifice for flow control is operated in a sampling train where the air enters the train at $P_{atm} = 692$ mm Hg. The head loss before the critical orifice is 28 mm Hg. What is the minimum vacuum gauge reading on the downstream side of the orifice that will ensure critical flow if the requirement is $P_d < P_{up} \times 0.45$? (T = 20 °C.)
3. If an orifice has a diameter of 0.45 mm and operates on a gas with T = 25 °C, and $P_{up} = 627$ mm Hg, compute the maximum mass flow rate in grams per second. Assume the discharge coefficient is 1.0. For critical flow rate in grams per sec:

$$\dot{m} = \frac{5.275 \times 10^{-3} C_d \times A_o \times P_{up}}{T^{\frac{1}{2}}}$$

where T is K, A_o is in mm², and P_{up} is in mm Hg.

4. A hypodermic needle is to maintain 200-mL/min flow of air at 20 °C and 1 atm. The head loss upstream of the orifice is 25 mm Hg. What diameter needle is theoretically required if $P_{up} = P_{atm} - 25$ mm Hg = 735 mm Hg and discharge coefficient is 0.97? The molecular weight of air is 28.96.
5. The limiting orifice of Problem 2 is calibrated at those specified conditions and found to control flow to 355 mL/min. The same orifice is then installed in another device where the P_{up} is 740 mm Hg and T = 30 °C. Calculate both the actual flow rate and the standard flow rate.

7 AIR POLLUTANT COLLECTION PRINCIPLES

7.1 INTRODUCTION

An important part of air quality maintenance is the measurement of airborne contaminant levels in the ambient (outdoor) air, in the work place (the industrial hygiene discipline), and in indoor residential areas or public areas. Air quality measurement is often accomplished by sophisticated, continuous air monitoring instruments. However, air sampling — the collection of a mass of pollutant for later analysis — also plays an important role, particularly for pollutants that are toxic at such low concentrations that a detectable mass of pollutant must be accumulated over time.

In the process of air monitoring or air sampling, two issues are of vital importance: accurate control of the flow rate and collection or detection of the pollutant in a manner that allows accurate qualitative and quantitative analysis. In air sampling, the volume of air sampled must also be accurately measured or calculated from the sampling rate and time, and the pollutant must be collected in the appropriate medium for removal from the air and storage. The analysis of the collected pollutant, including proper preservation of the sample, is covered in great detail in numerous references.[7,11,27,28,38]

In this chapter, the primary topic is proper sampling equipment and techniques. Many of the issues covered in Chapters 2 through 6 apply. Air sampling flow rate monitoring is often accomplished by rotameters (Section 4.6), orifice

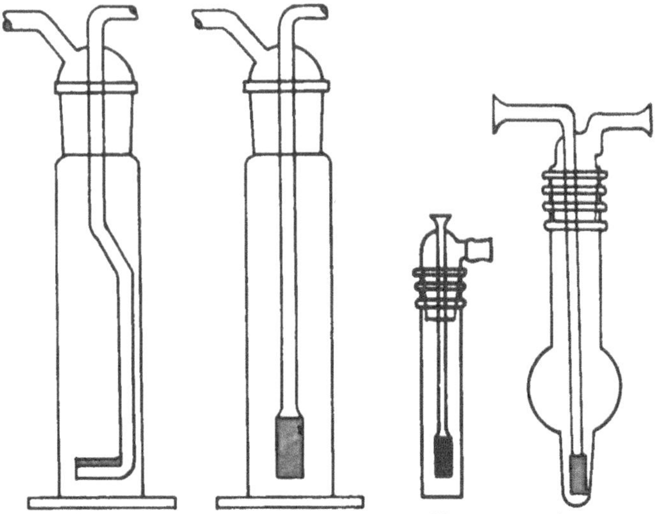

Figure 7.1. Absorption devices (Reference 15).

meters (Section 4.3), mass flow meters (Section 4.7), or capillary tubes (Section 4.5). Flow control by critical orifices (Section 6.2), or electronic or differential pressure flow controllers (Sections 6.3 and 6.4), is standard practice. Often, sampling rates are set or checked through use of a bubble meter calibration system (Section 3.5).

Calibration of the sampling or monitoring flow rate must be performed under conditions identical to those of the actual air sampling process — the pollutant collection or detection medium must be assembled with the air sampling train when flow rates are checked. This procedure is known as *in situ* calibration.

The remainder of this chapter provides a brief review of pollutant collection techniques used in air sampling. The following chapters give specific details on calibration and use of some air sampling devices.

7.2 ABSORPTION OF GASES

Before continuous air monitoring instruments were developed, gas–liquid absorption was the technique used for nearly all gaseous air pollutant sampling. Absorption of air pollutants in a liquid and later analysis is still the "EPA Reference Method" for sulfur dioxide in the air and is an approved method for many low molecular weight compounds.

Absorption is the process of transferring a gaseous compound into a uniform distribution in a liquid or solid. In air sampling work, the absorbent is usually a liquid. There are two types of absorption: physical and chemical. Physical absorption involves the dissolving of the pollutant in a liquid. The solubility of a particular air pollutant depends on the absorbent, its temperature, and the partial pressure of the pollutant in the air. Efficiency of collection of the gas of interest as the sampled air passes through the absorbent is usually fairly low for physical absorption.

Chemical absorption uses a liquid that reacts with the pollutant to produce a unique, stable, nonvolatile, and easily detected product in the liquid phase. An example would be the measurement of ammonia gas by passing the air containing the gas through an acidic solution, such as boric acid. The nonvolatile borate ion $H_2BO_3^-$ thus produced can then be determined quantitively.

Absorption sampling devices are normally glass containers fitted with leak-tight seals, a vacuum outlet, and an inlet tube with its end submerged. The inlet tube is fitted with a fritted glass end with 50-μ or smaller pores, or with an impinger inlet. (See Figure 7.1.)

7.3 ADSORPTION OF GASES

Adsorption involves the collection of gases, liquids, or solutes on a solid *surface*. Usually, the surface is a porous solid sorbent to which gas molecules are attracted. For maximum adsorption in air sampling, granules of a porous material (e.g., charcoal) are used to provide very large surface areas. The sorbent has a crystal lattice structure — atoms or molecules arranged in a regular pattern. At the surface, valence forces that link the lattice atoms are unsatisfied, so the surface attracts gas molecules (the adsorbate). Van der Waals' nonideal gas molecular interaction forces also play an important role in physical adsorption. Most air sampling work involves this physical adsorption; chemical adsorption (involving a reaction) is limited at low temperatures.

Because the same type of forces that cause condensation is involved in physical adsorption, the quantity of any gas adsorbed under a selected set of conditions is directly related to the ease of condensation of the gas. Thus, the boiling point of a compound provides a good indication of the ability to adsorb it — the higher the boiling point (or the less volatile a compound is), the more gas will be adsorbed.

Physical adsorption is generally reversed (desorption) by: (1) high temperature, or by (2) solvent rinsing. Both of these techniques are used in air sampling adsorption.[11,34,35]

The applications of adsorption/desorption to air sampling are given in Table 7.1.

Table 7.1 Applications of Adsorption/Desporption to Air Sampling

Technique	Application	Example
Thermal desorption	Volatile organic sampling	EPA Method TO-1 for benzene
Solvent desorption	Industrial hygiene sampling	NIOSH Method 1501 for toluene
Gas chromatography	Most EPA and NIOSH organic vapor analysis methods	Separation of mix of organic vapors for detection by FID or MS

The quantity of a gaseous pollutant adsorbed by a given quantity of adsorbent depends on:

1. Concentration in the air over the surface — Higher concentration or partial pressure of the gas of interest in the air will lead to greater adsorption; lowered concentration will cause desorption. 100% collection efficiency in an adsorption air sampling system is not technically possible, as some equilibrium point will be reached between the concentration in the air and the quantity adsorbed. However, in air sampling by adsorption, efficiencies are very high, and the collection is generally regarded to be quantitative.
2. Surface area of the adsorbent — The best adsorbents are granules of porous materials like silica gel or coconut shell charcoal, which provide very large surface areas in a small volume and do not restrict the flow of the gas drawn through the granule bed. A granular adsorbent can take up to 40% of its weight in adsorbate. Activated charcoal (charcoal heated to 350 to 1000 °C in a vacuum or inert gas to distill out impurities and open pores) has long been utilized, though polymeric granule adsorbents that have better desorption characteristics are now more common.
3. Temperature — Moderately high temperatures will reduce adsorption efficiency by shifting the equilibrium point toward a higher air concentration. Higher temperatures will nearly completely desorb the adsorbate in some cases. Room temperatures or lower are best for air sampling by adsorption.
4. Other gases competing for adsorption sites — Other compounds in the sampled air may be preferentially adsorbed, filling some or all of the adsorption sites. Complete fouling of the adsorbent with other compounds may occur. Some air sampling systems utilize multiple beds of different adsorbents in series to collect many different compounds in air without contamination problems.[36]

5. Properties of the adsorbate (size, polarity, etc.) — Large-molecule gaseous compounds are typically more easily adsorbed (fluorocarbons are an exception). Light (low molecular weight) compounds are difficult to physically adsorb. Light compounds like formaldehyde, ammonia, and hydrogen chloride may be adsorbed chemically or, more frequently, are sampled by absorption. The degree of polarity of the adsorbent plays an important role in the ability to collect adsorbates. Activated carbon is composed of nonpolar atoms with little electrostatic polarity, and is effective for collecting many low-polarity organic compounds. Adsorbents made of siliceous compounds (e.g., molecular sieves) tend to be polar and have a much greater affinity for polar gases. They are frequently used in gas chromatographic separation columns, but have the drawback of attracting water molecules, which are strongly polar.

As noted above, charcoal is often used for sampling for organic compounds in air. Recently, nonpolar, hydrophobic organic polymers have been used as adsorbents in air sampling for organic compounds. Two of the most important are TENAX® and Amberlite XAD-2® resin. Table 7.2 lists the chemical nature and application for these two organic polymeric adsorbents.

Table 7.2 Organic Polymeric Adsorbents

Common name	Chemical nature	Applications
TENAX-GC®	Poly (2,6-diphenylphenylene oxide) — volatile	Organic sampling
Amberlite XAD-2®	Polyaromatic resin — semivolatile	Organic sampling

(TENAX-GC is a registered trademark of Enka Research Institute; Amberlite XAD-2 is a registered trademark of Rohm & Haas Co.)

7.4 OTHER GASEOUS POLLUTANT SAMPLING METHODS

Some newer and important air sampling methods include condensation or cryogenic sampling and whole air sampling.

Cryogenic Sampling

The collection of air pollutants by condensation in a cryogenic trap is a preferred method for volatile organic compounds present in the air at very low

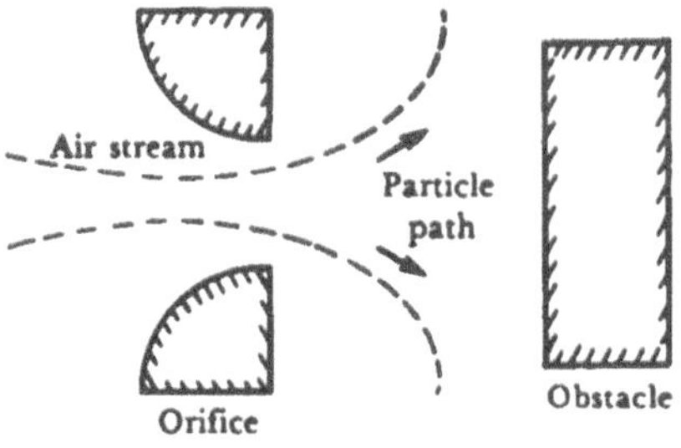

Figure 7.2. Inertial or impaction collection (Reference 15).

concentrations. Condensation in the cold trap allows preconcentration for easier analysis by gas chromatography/mass spectroscopy (GC/MS).

Whole Air Sampling

Collection of air samples in inert glass bulbs, tedlar or teflon bags, gas-tight syringes, or evacuated polished stainless steel canisters are examples of whole air sampling. Air contaminants present at very low levels may require concentration before analysis.

7.5 INERTIAL COLLECTION OF PARTICULATE MATTER

Collection of suspended particulate matter in the air utilizes principles very different from those in the collection of gases. Primarily, the size or inertia of the solid particle or liquid droplet is the property capitalized upon for collection. Inertial collectors use the principle that particles in a gas stream will tend to be deflected less than the nearby gas when the gas stream is subjected to a sudden change in direction. If the velocity is large enough and the direction change around a solid surface severe enough, the particles' inertia will cause them to be thrown against or impacted on the surface.

Particles moving through a gas are characterized by their *aerodynamic diameter*. Aerodynamic diameter takes into account the shape, roughness, and aerodynamic drag of a particle. A particle that moves through a gas at the same terminal velocity as a smooth spherical particle of a certain diameter is said to have that *aerodynamic diameter*.

The velocity of the particle-carrying air stream and the severity of the direction change in an inertial collector can be designed to cause most of the particles with a desired aerodynamic diameter range to strike the impaction surface (see Figure 7.2). The collection surface in an inertial air sampler may be a glass fiber mat (filter paper is used, but does not actually filter the gas),

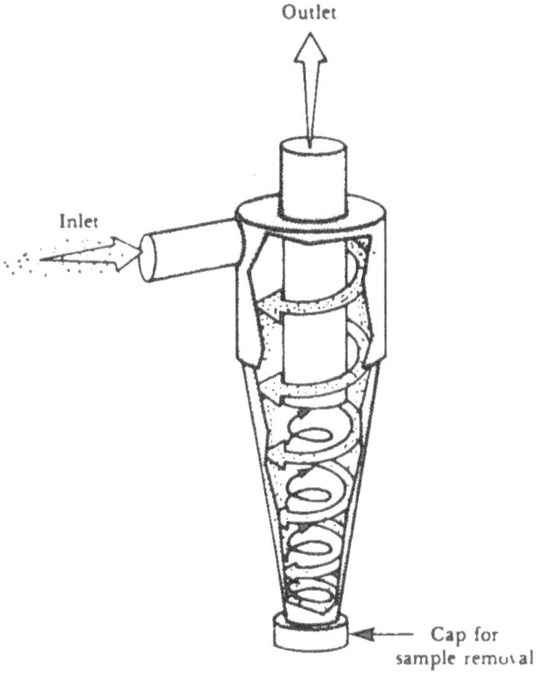

Figure 7.3. Cyclone collector for particulates (Reference 15).

a solid surface coated with grease or oil, or simply a hard surface oriented nearly parallel to the gas flow (as in the cyclone collector depicted in Figure 7.3). The collection surface is typically weighed before and after sampling to determine the amount of particulate matter collected. For a well-designed inertial collector, there is a narrow range between the smallest-sized (aerodynamic diameter) particles collected and the size above which all particles are collected. The aerodynamic diameter on such a distribution which is collected with 50% efficiency is usually used to describe the collector and is called the "50% cut point". In Figure 7.4, the 50% cut point is 10-μm diameter particles.

7.6 FILTRATION FOR PARTICULATE MATTER

Collection of suspended particles from the air by filtration is by far the most popular technique, particularly when no breakdown of the sampled particle distribution by size is desired.[26] Filters can be used to assess total dust loading, or with an inertial precollector, to measure a subset of smaller-diameter particles. Exposed filters can also be analyzed microscopically or chemically.

Materials used in air sampling filters include glass or quartz fiber, cellulose, organic membranes, or polyurethane foam. Each type of filter has certain

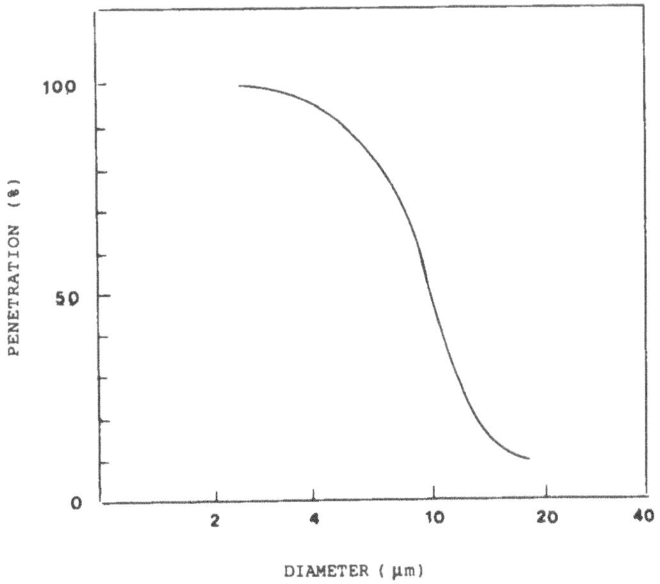

Figure 7.4. Inertial collection efficiency.

applications in air sampling and most are available in numerous configurations, from 8 × 10" sheets to 25-mm circles loaded into sampling cassettes (see Figure 7.5). The three main categories of filter materials are discussed below.

Glass Fiber Filters

Glass fiber filters have had the longest history of application to air sampling. This type of filter is specified by the EPA for use in total suspended particulate matter samplers in the National Air Sampling Network, and has a number of advantages in air sampling. Typically manufactured by combining finely spun borosilicate glass fibers with a binder and compressing into a thin mat, the filters exhibit low water uptake, low head loss, and high capture efficiency for particles larger than 0.3 µm. Glass fibers are resistant to most corrosive gases and can operate at temperatures up to 540 °C, which makes such filters very appropriate for particulate matter emission sampling by EPA Methods 5 and 17.

The finished filters can be heat-treated to remove organic binder material, rendering them even more inert, but somewhat more brittle. Various manufacturers produce glass fiber filters that are especially low in trace metals, such as zinc or iron, or have controlled sulfur content or pH.

All glass fiber filters capture particles at various depths in the mat of fibers. Glass fiber filters are less efficient for very small particles, but their relatively low cost and high load capacity make them appropriate for many air sampling tasks.

High-purity quartz fiber filters are specified by EPA for PM10 sampling.

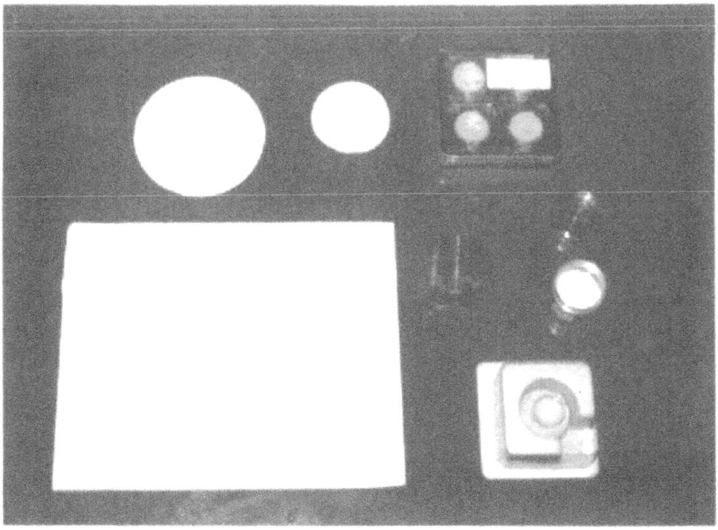

Figure 7.5. Air sampling filter configurations.

Cellulose Filters

Cellulose fiber filters are made of purified cellulose pulp. Although cellulose fibers are hydrophilic (i.e., take up water easily), this type of filter has some advantages that make it useful to some air sampling applications: low head loss, low level of metallic contaminants, and the ability to be easily ashed for analysis of the collected matter.

Organic Membrane Filters

Membrane filters are commonly thought of as surface-collecting filters. There are a number of types of membrane filters, falling in two main categories. One type is formed as a thin polycarbonate or polyester membrane, which is later etched to produce standard-size pores of the desired density. The other type is produced by dissolving pure cellulose nitrate and cellulose diacetate or other polymers in organic solvents, casting on a flat surface, and evaporating the solvent. In some cases, particularly with teflon filter material, the membrane is stretched to produce the desired pore size.

Membrane filters have the advantage of very uniform pore size, allowing efficiency for a given particle size to be accurately predicted. Membranes have the added advantage of collection on a flat surface, allowing for microscopic examination of the sample. Most membrane materials are hydrophobic and resistant to chemical attack. Mixed esters of cellulose membrane filters are free of contaminants and have extremely low portions of water-extractable material, minimizing weight loss problems with gravimetric analyses.

The major drawbacks of membrane filters are their large head loss, brittleness, and inability to withstand high temperatures. Membrane filters have found most frequent air sampling applications in small-diameter filters used in dichotomous samplers, asbestos samplers, and other relatively low-flow samplers. See ACGIH[40] for further discussion of filter media.

Collection Mechanisms

Air filtration involves a number of collection mechanisms; filters for removal of suspended particles from air do not function primarily by sieving, but rather by diffusion, direct interception, and/or inertial collection.

Diffusion

As the particle-laden air passes through the filter, the particles diffuse from the high-concentration areas in the air stream to the areas near the filter material where the concentration of particles is near zero. The amount of collection by the diffusion mechanism depends on gas velocity, filter thickness, particle size, filter pore size or interfiber distance, and particle concentration.

Direct Interception

When particles that are carried by the air stream through the pores in the filter pass within touching distance of filter material or fibers, they are collected by direct interception. Small pore size and large particle size lead to increased collection by interception.

Inertial Collection

As in inertial collectors, when the air stream is sharply turned around the filter material or fibers, the particles' inertia causes them to come into contact with the surface. High velocities, large particle size, and small pore size, give higher collection by inertial mechanisms.

Air sampling filters are available with a wide range of pore sizes. Filters do not collect 100% of the particles of all sizes in the air stream; large percentages of small-diameter particles obviously pass through some filters. However, appropriately selected filters typically do have very high and repeatable collection efficiencies for the same sampling conditions and are generally treated as 100%-efficient collectors. Very fine particles require filters with very small pore size and, consequently, a large pressure drop across the filter. Cadle[41] gives a thorough discussion of filtration theory.

7.7 OTHER PARTICULATE POLLUTANT SAMPLING METHODS

Microscopic identification or counting of particles collected on a filter or impaction surface has been an important particle assessment technique for many years. Phase contrast microscopy and electron microscopy are particularly important techniques for asbestos monitoring.

Electrostatic precipitation of particles out of the air has been employed as a collection technique in air samplers for indoor air quality work.

In "clean room" technology, the number of particles is more significant than the mass of particulate matter per volume of air. In microchip manufacture, for example, the concern is with the number of particles that can fall onto and damage the microcircuits during manufacturing. Particle counters, utilizing laser technology and detecting each single particle by the light it scatters, are widely used, particularly in the electronics industry.

8 AIR SAMPLING METHODS FOR GASEOUS POLLUTANTS

8.1 INTRODUCTION

For traditional or "EPA criteria" air pollutants, there are many continuous or real-time air monitors that will accurately detect air pollutants at concentrations in the range of a few parts per billion (ppb). For example, the National Ambient Air Quality Standard for Ozone is 120 ppb, and continuous ozone monitors are required to reliably measure concentrations as low as 10 ppbv. However, many ambient air toxics regulations and many indoor air quality regulations require detection of pollutants at concentrations in the single ppb range, or even in the parts per trillion (ppt) range (see Table 8.1). For such low concentrations, and for many air pollutants at substantially higher levels, collection of a measurable mass of the gaseous pollutant from the air, followed by later analysis, is the preferred approach.

Table 8.1 Examples of Vermont Hazard Limit Values

1,1,2,2-Tetrachloroethane	2.48 ppt
Hexachlorobenzene	0.18 ppt
Cadmium	0.12 ppt
2,3,7,8-Tetrachlorodibenzodioxin	1.52×10^{-6} ppt

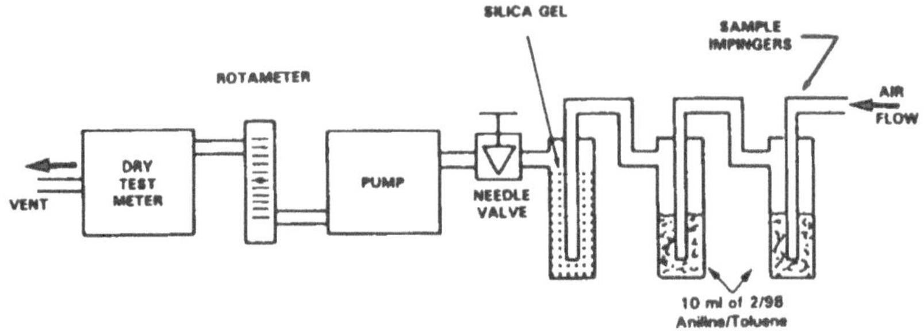

Figure 8.1. Typical sampling system for phosgene (Reference 11).

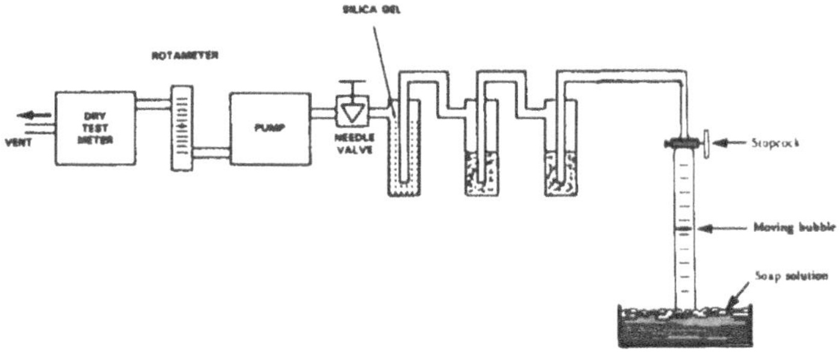

Figure 8.2. Calibration of phosgene sampler.

This chapter presents some examples of applications of sampling and pollutant collection methods discussed in Chapter 7 to gaseous air pollutant measurements.

8.2 IMPINGERS (ABSORBERS)

Hydrogen chloride, ammonia, aldehydes and ketones, phosgene, and cresols are examples of air pollutants best sampled by liquid absorption or impingement methods. In several EPA-developed ambient air sampling methods, midget impingers (25-mL glass bubblers) are installed in a very simple sampling train as shown in Figure 8.1. A specified quantity of air is drawn through the impingers at an established flow rate. The liquid in the absorber reacts with the compound of interest to form a unique and stable substance. Analytical methods such as high-pressure liquid chromatography (HPLC) are used to quantify the amount of the pollutant of interest collected.

As an example, Method TO6,[11] for measuring phosgene in ambient air, specifies two midget impingers in series with 10 mL sampling reagent in each, calibration with a soap-bubble meter or wet test meter to a flow rate not greater than 1000 mL/min, and total air volumes of 50 L or less (see Figure 8.2).

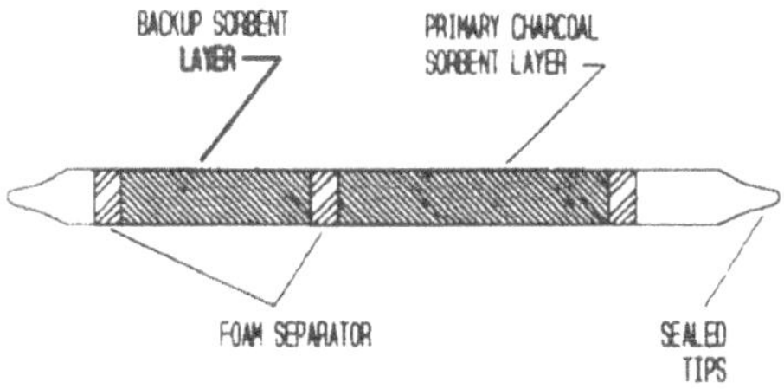

Figure 8.3. Industrial hygiene charcoal adsorber.

8.3 ADSORBERS

Solvent Desorption Tubes

There are numerous applications of adsorption in air sampling. Industrial hygiene studies for volatile organic compounds in the work place have utilized charcoal adsorption tubes and solvent desorption with gas chromatography/flame ionization analysis techniques for many years (see Figure 8.3).

As an example, worker exposure to benzene is a frequently expressed concern. NIOSH Method 1501 for benzene can make use of a portable system for sampling that includes a personal flow-controlled pump, light enough for a worker to carry all day, and a sorbent tube of the type shown in Figure 8.3. The sampling inlet is located near the worker's breathing area at all times, and may even be used inside a respirator. The system samples at a very low flow rate of 20 mL/min for 8 hr and, in that time, samples nearly 10 L of air (see Figure 8.4).

Colorimetric Tubes

Another type of adsorption tube that has value for screening an area to determine best placement of more precise sampling units is the colorimetric tube, or indicating detector tube (Figure 8.5). This type of tube is filled with a coated granular material. The coating in each type of tube is designed to react colorimetrically with a single pollutant and produce a stain inside the tube. In some applications the length of the stain, and in others the intensity of the color, is proportional to the concentration of the pollutant in a specified amount of air sampled. The air mover can be a calibrated low flow rate battery-powered pump, a piston pump, or a hand-operated bellows.

Thermal Desorption Tubes

EPA Methods TO1 and TO2[11] specify the use of TENAX® and molecular sieve, respectively, to trap volatile organic compounds from sampled air. The

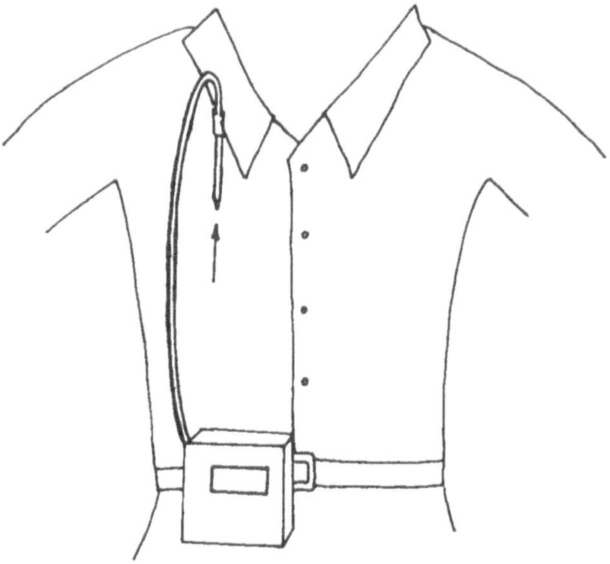

Figure 8.4. Personal sampling.

exposed sample tubes are later installed in a thermal desorption unit, where they are rapidly heated to desorb the collected compounds. A carrier gas carries the released pollutants into a gas chromatograph with mass spectroscopy (GC/MS) detector. The sampling system generally used is identical to that used for solvent desorption tubes. An alternative method, useful for the same volatile compounds, employs a three-bed sample adsorption tube, in which pollutants of various molecular weights or boiling points are collected in successive adsorption beds (Figure 8.6). This system also employs thermal desorption to release the compounds to the GC/MS detector for analysis. Reaction of collected compounds during adsorption or desorption and the consequent production of artifact species is sometimes a problem with this type of sampling.

Adsorber Flow Control and Sampling Train Calibration

In many adsorber sampling applications, personal battery-powered, flow-controlled pumps are used. The complete sampling system, with a spare adsorption tube installed, must be calibrated, typically with a bubble meter. In the calibration process, the pump is adjusted to the desired flow rate (see Figure 8.7).

In other applications, critical orifice flow control is employed, sometimes with several adsorption tubes in parallel collecting replicate samples. Another method for multiple parallel samples uses electronic flow controllers which both maintain set flow rates and record total volume sampled. In the example shown in Figure 8.8, each flow controller is set to a different rate. Critical orifice or flow controller flow rates are checked, with the complete sampling train assembled, by means of a bubble meter.

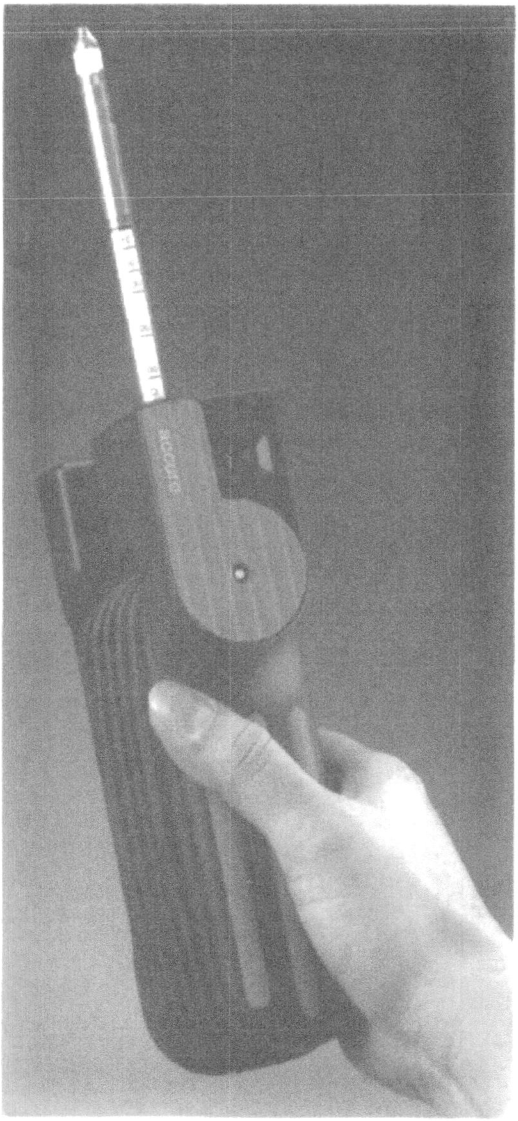

Figure 8.5. Colorimetric tube. (Courtesy of National Draeger, Pittsburgh, Pennsylvania.)

Emission Isolation Flux Chamber Sampling

An interesting application of adsorption tube sampling for volatile organic compounds is the isolation flux chamber, used to measure emission rates of pollutants from landfill outgassing or hazardous waste disposal sites[24] (see Figure 8.9).

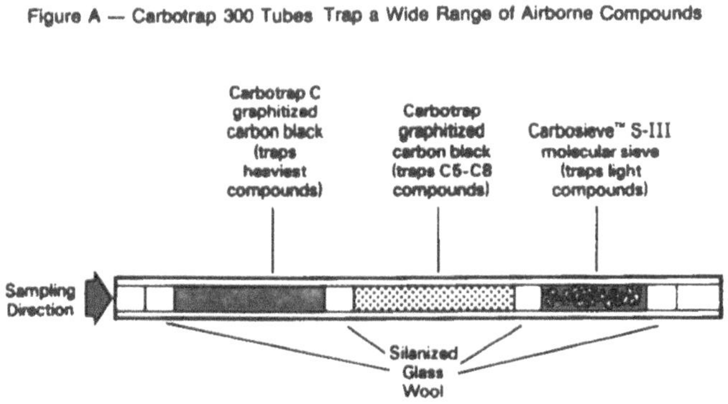

Figure 8.6. Three-bed adsorption tube Supelco Tarbotrap 300. (Courtesy Supelco, Inc., Bellefonte, Pennsylvania.)

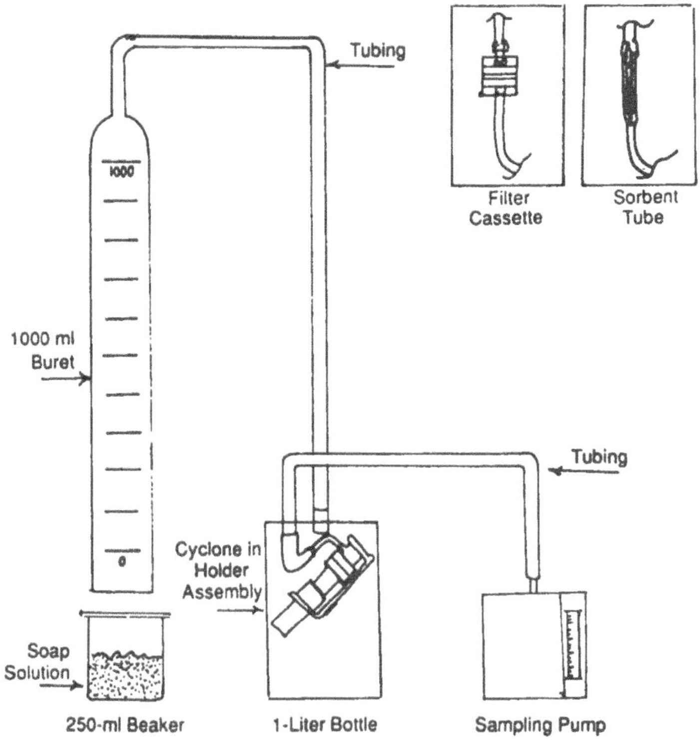

Figure 8.7. Calibration of personal sampler. From *OSHA Industrial Hygiene Reference Manual*, Reference 35.

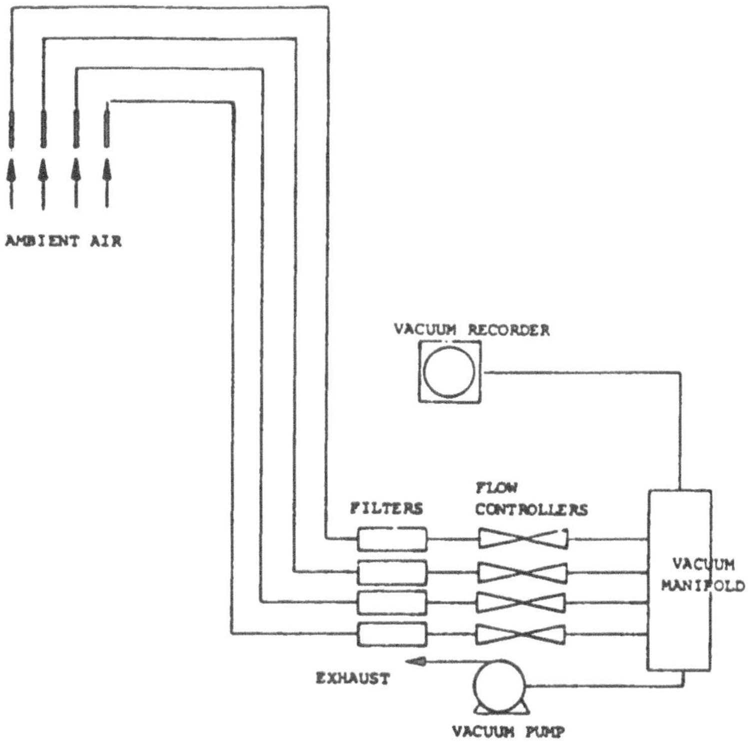

Figure 8.8. Flow diagram for toxic air monitoring system (TAMS) air sampler.

8.4 CRYOGENIC SYSTEMS

EPA Methods TO3 and TO12 specify cryogenic trapping or preconcentration for sampling volatile organic compounds in the air.[37] Liquid oxygen or argon is used to cool a U-shaped piece of tubing filled with glass beads. Air is drawn through the trap, where volatiles condense out. After sampling is complete, the trap is rapidly heated and the pollutants revaporize and are swept by carrier gas to the analyzer. A drawback of cryogenic trapping is the potential for unwanted reactions of water-soluble condensates with each other, obviating identification of the individual compounds in the air sample. See Figures 8.10 and 8.11.

8.5 WHOLE AIR SAMPLERS

Whole air samples can be collected by means of a displacement bottle (Section 3.4), a bag inflation sampler, a sampling syringe, or an evacuated canister or flask system.

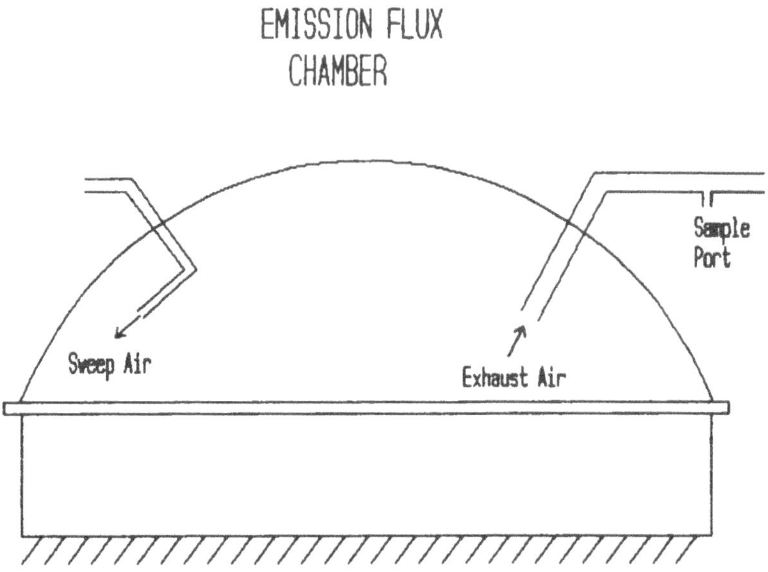

Figure 8.9. Emission isolation flux chamber.

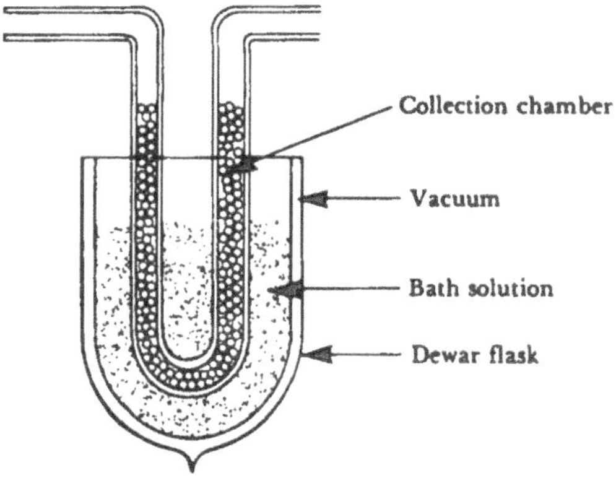

Figure 8.10. Cryotrap sampling system (Reference 11).

For some nonreactive compounds, sampling into a tedlar or teflon bag, as in Figure 8.12 is simple and effective. The sample can then be transported to the laboratory for analysis. Soil gas monitoring is one application of this technique.

Large gas-tight syringes make convenient grab sampling devices (Figure 8.13).

EPA Method TO14,[11] for determination of volatile organic compounds in ambient air, requires the use of polished or SUMMA® passivated stainless steel

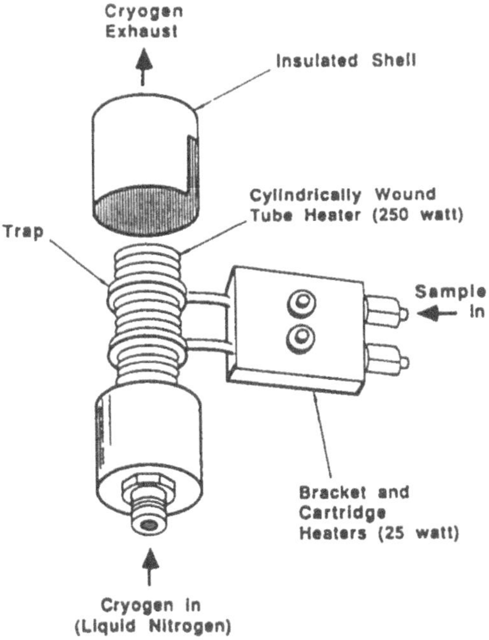

Figure 8.11. Cryogenic trapping unit (Reference 15).

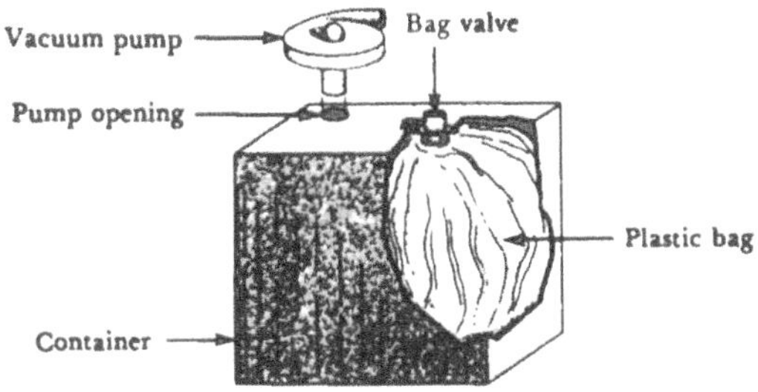

Figure 8.12. Bag inflation sampler (Reference 15).

canisters to collect the sample. The system can use either an evacuated canister which can fill without a pump, or it can employ a pump to collect a pressurized sample. Analysis of the sample, per Method TO14, requires cryogenic preconcentration and gas chromatography with a choice of detectors (Figures 8.14 and 8.15).

The EPA Methods for Determination of Toxic Organic Compounds in Air, as published in EPA-600/4-84-041,[11] and addenda are listed in Appendix B.

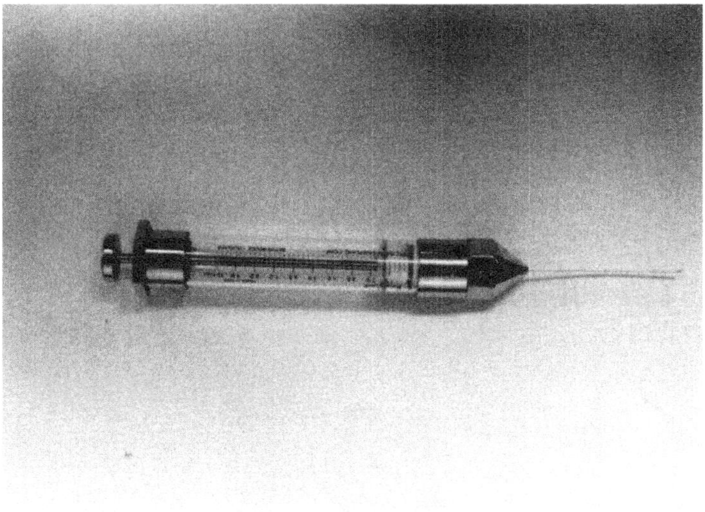

Figure 8.13. Air sampling syringe.

Figure 8.14. Canister sampler. (Courtesy of Graseby-Andersen, Atlanta, Georgia.)

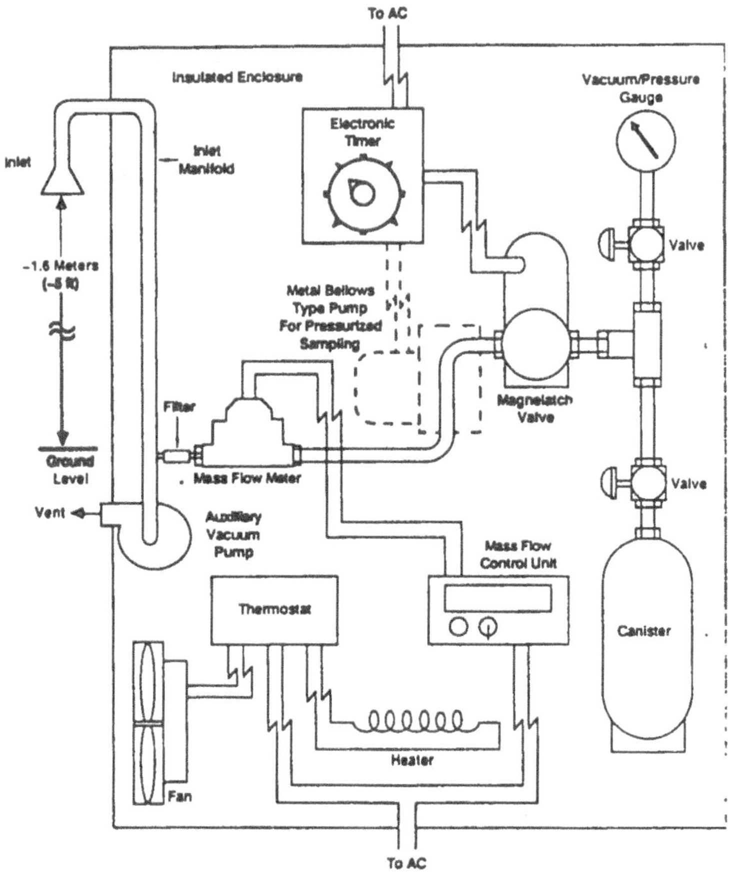

Figure 8.15. Canister sampling set-up (Reference 11).

9 AIR SAMPLING METHODS FOR PARTICULATE MATTER BY FILTRATION

9.1 INTRODUCTION

Measurement of suspended particulate matter in the air is likely the oldest form of air quality sampling. High-volume samplers, PM10 samplers, dichotomous samplers, polyurethane foam samplers, industrial hygiene personal samplers, and asbestos samplers all employ filtration sampling and require flow monitoring and sampler air volume measurement. The basic applications of filtration theory to particulate air sampling are reviewed in this chapter. Sections 9.7 and 9.8 give specific details on calibration techniques.

9.2 HIGH-VOLUME SAMPLER

The high-volume sampler (Figure 9.1) was developed in 1948[29] in essentially the same form as used today. The "high-vol" sampler draws a large, accurately known volume of air over a 24-hr period through a glass fiber or membrane filter. Particles in the air with diameters between 0.3 and ~100 μm are collected on the filter, which is weighed before and after sampling, and the average concentration over the sampling period is calculated. High-volume sampling and modern adaptations are further discussed later in this chapter.

Today, the high-volume sampler is used primarily to provide samples of particulate matter for laboratory analysis for metals, organic particles, and sulfate and nitrate compounds. In addition, high-vol samplers (Figure 9.2) are

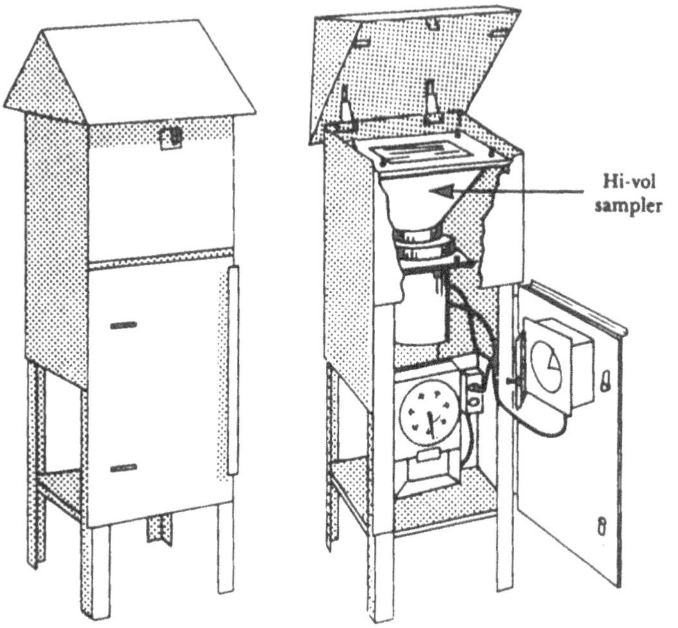

Figure 9.1. High-volume sampler (Reference 15).

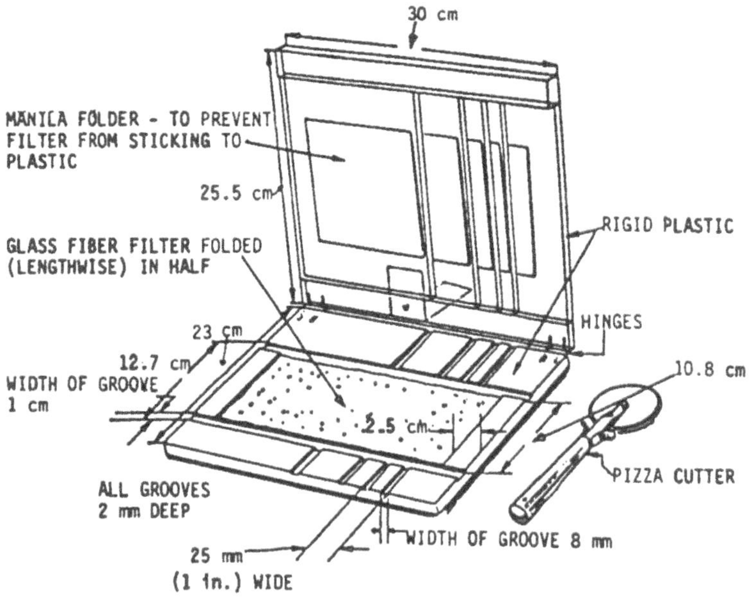

Figure 9.2. High-volume filter cut for metals, organics, and ion analysis.

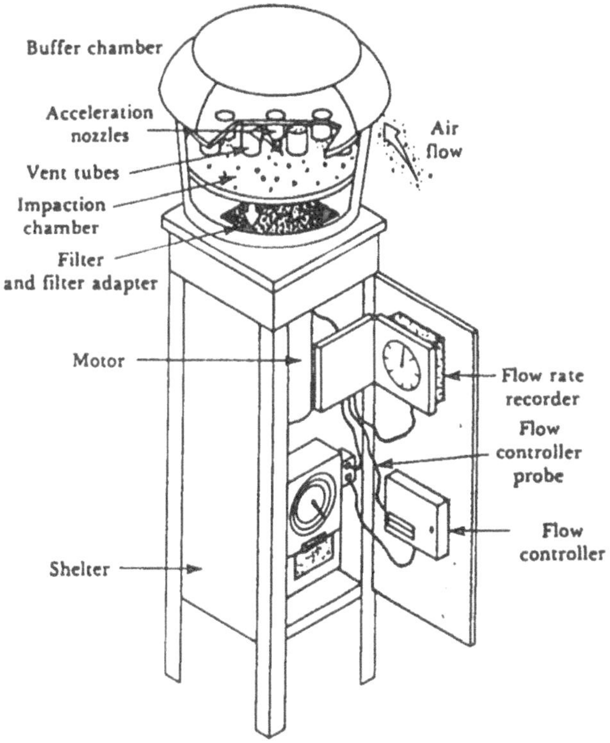

Figure 9.3. High-volume sampler with PM10 inlet (Reference 15).

run in some locations to measure concentration of suspended particulate matter to continue a historical record spanning 30 years or more.

In the 40+ years since the development of the high-vol sampler, there have been many refinements to suspended particulate matter sampling, including changes in filter materials and sampler inlet design. However, the fundamental concepts, flow measurement methods, and flow control techniques remain basically the same in today's PM10 samplers and dichotomous samplers. Nearly all ambient (outdoor) particulate matter samplers utilize a high flow rate vacuum pump (0.59 cfm or 16.7 lpm for dichotomous samplers; up to 90 cfm or 2.55 m^3/min for high-vol samplers) driven by an electric motor.

9.3 PM10 SAMPLER

Measurement in the U.S. of the total mass loading of particles in the air, of concern as air pollution, is now largely done by modifications to the standard high-vol samplers called "PM10" samplers (Figures 9.3 and 9.40). PM10 stands for particulate matter less than or equal to 10 μm in diameter. The EPA decision, finalized in 1987, to use PM10 samplers as the primary ambient

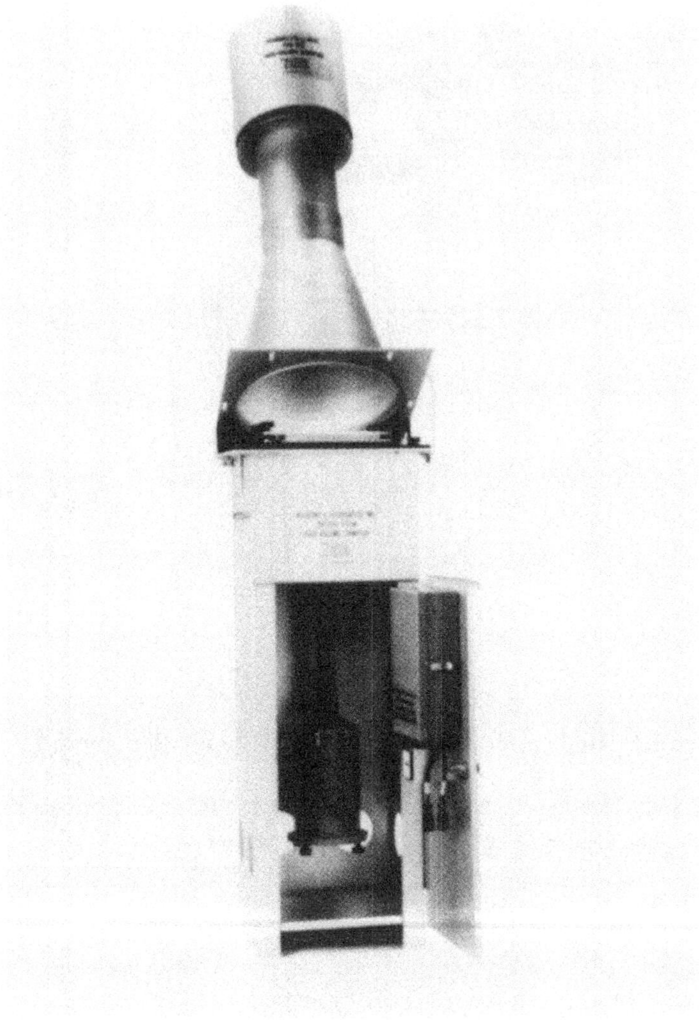

Figure 9.4. PM10 sampler. Courtesy of Wedding & Associates, Ft. Collins, Colorado.

particulate matter monitor results from the finding that only particles with diameters less than 10 µm are of concern for respiratory air pollutant damage to humans.

PM10 samplers function basically in the same fashion as high-volume samplers, and require the same or more care in flow rate and volume measurement, but arrange to remove the larger particles before the air reaches the filter through variously designed impaction or settling chamber inlets. Flow control by electronic sensors (Section 6.3), differential pressure controllers (Section 6.4), or critical orifices (Section 6.2) is employed.

9.4 POLYURETHANE FOAM SAMPLER

Significant quantities of both vapor and aerosol forms of a semivolatile organic compound may be present in air if the compound's vapor pressure is moderately high. In addition, some gases may adsorb onto other particles in the air. Further, during traditional filtration sampling, some collected particulate material with relatively high vapor pressure may evaporate as more air is passed through the sample already collected on the filter.

To overcome these problems in sampling semivolatile organic compounds in air, polyurethane foam (PUF) samplers are used. The PUF sampler is a modification of a high-volume sampler in which a quartz fiber filter is followed by a 6-cm plug of foam (see Figure 9.5). Polyether-type polyurethane foam (density 0.0225 g/cm^3) will efficiently collect vapor and particulate forms of pesticides, polychlorodiphenyls (PCBs), dioxins, and some polynuclear aromatic hydrocarbons (PAHs). The compounds of interest are collected in solid or droplet form on the filter and in vapor form on the foam. In some applications, an adsorbent resin (XAD-2®) is added behind the foam plug to further enhance the capture of more volatile compounds.

9.5 DICHOTOMOUS SAMPLER OR VIRTUAL IMPACTOR

Another sampler that collects particulate matter is the virtual impactor or dichotomous sampler (Figure 9.6). As the names imply, the technique uses impaction of particles into a void (virtual surface) and, by so doing, allows separation of the particles into two size ranges. Since this type of sampler normally utilizes a PM10 inlet, the two size ranges are typically particles between 10 and 2.5 μm in diameter, and those less than 2.5 μm. Dichotomous samplers use smaller flow rates than other PM10 samplers and smaller membrane filters. Precise flow control is essential to their operation; flow controllers like those discussed in Chapter 6 are used. Details of dichotomous sampler operation are reviewed by Loo et al.[39]

9.6 INDOOR PARTICLE MONITORS

Most of the particle samplers discussed above tend to be too noisy or obtrusive for use indoors. Smaller flow rate and physically smaller devices have

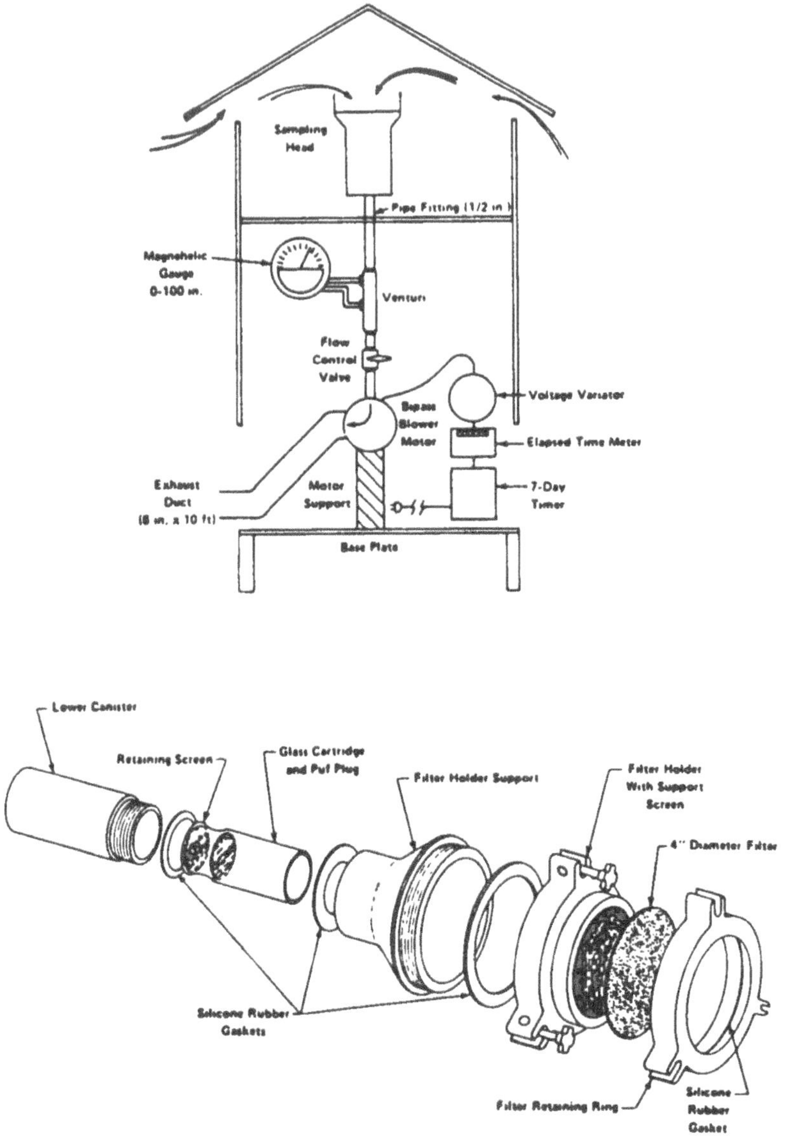

Figure 9.5. PUF sampler (Reference 11).

been used for industrial hygiene studies and residential indoor measurements.[9,21] See Figures 9.7 through 9.11.

As with personal gaseous pollutant samplers, flow rate calibration for particle sampling systems is often done by means of bubble meters or mass flow meters (Chapters 3 and 4). Flow-controlled, battery-powered personal pumps are frequently employed for sampling (Figure 9.12).

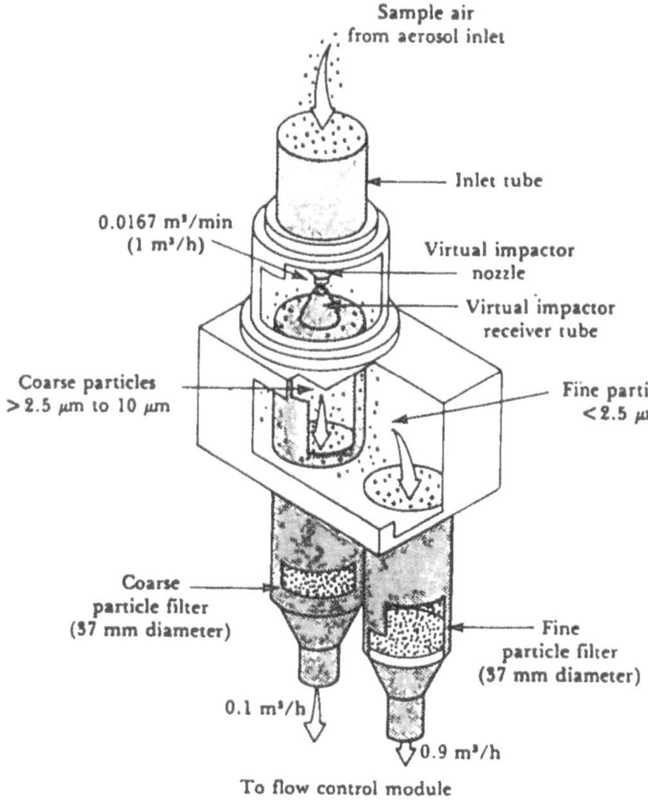

Figure 9.6a. Dichotomous sampler flow diagram. (Courtesy of Graseby-Andersen, Atlanta, Georgia.)

9.7 FLOW CALIBRATION FOR HIGH-VOLUME SAMPLERS

Particulate matter sampling requires careful filter handling before and after sampling to avoid moisture and weighing errors. Equally important to accurate suspended particulate matter measurement, however, are accurate flow control, flow rate, time, and sampled volume measurement. For many years, a two-step flow calibration procedure has been used in high-volume sampling, which, with some modifications, is still valid today.[42] The procedure is also applicable to PUF and PM10 samplers and provides good insight into the general process of air sampling calibrations.

Calibration Step 1

A laboratory-based positive displacement meter or Roots meter (see Section 3.9) is used to develop a calibration curve for a sharp-edged orifice (Figure 4.3).

Figure 9.6b. Dichotomous Sampler.

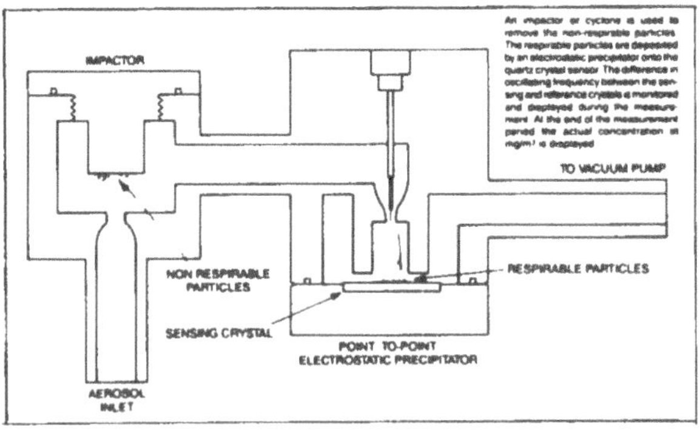

Figure 9.7. TSI electrostatic precipitator. (Courtesy TSI, Inc.)

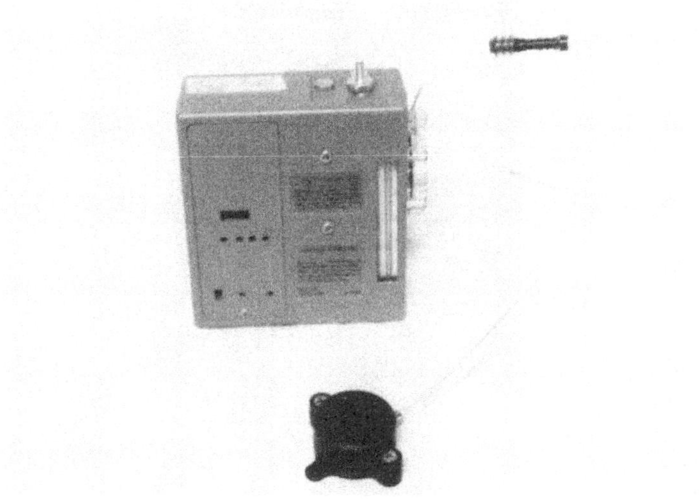

Figure 9.8. Personal exposure monitor.

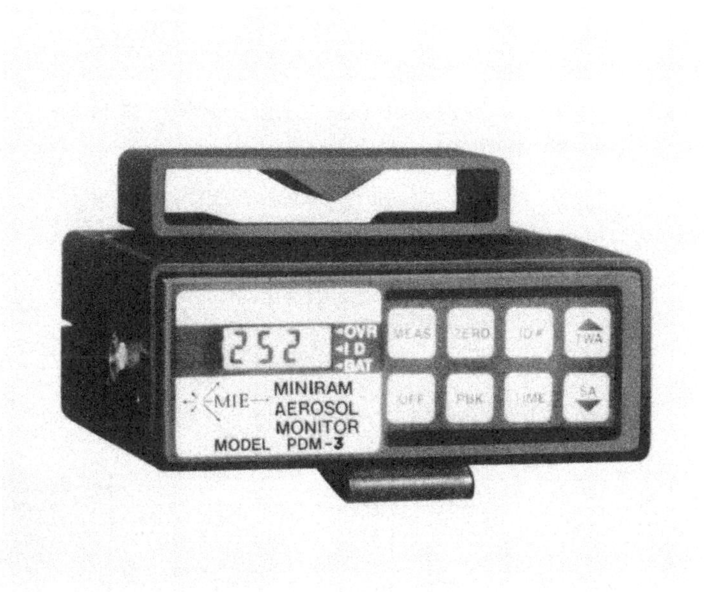

Figure 9.9. Inhalable particulate personal monitor. (Courtesy MIE, Billerica, Massachusetts.)

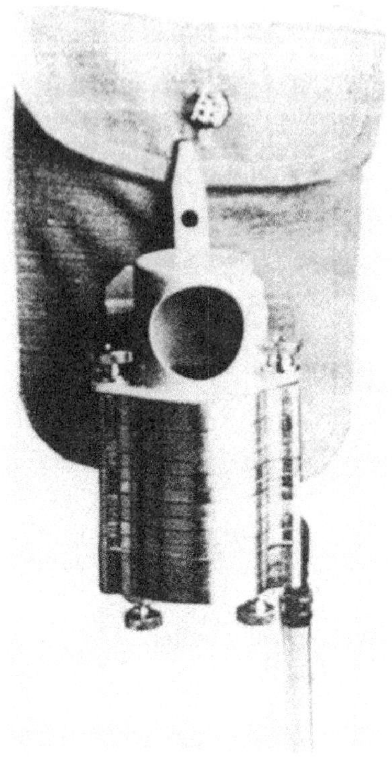

Figure 9.10. Marple personal cascade impactor. (Courtesy Sierra Instruments.)

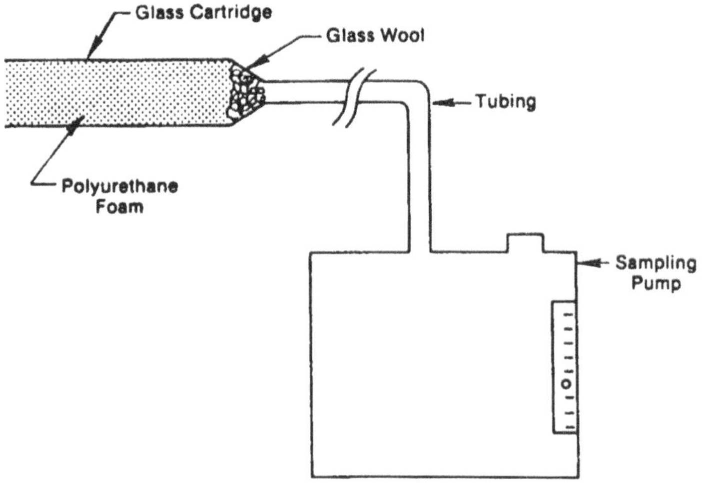

Figure 9.11. Portable PUF sampler.

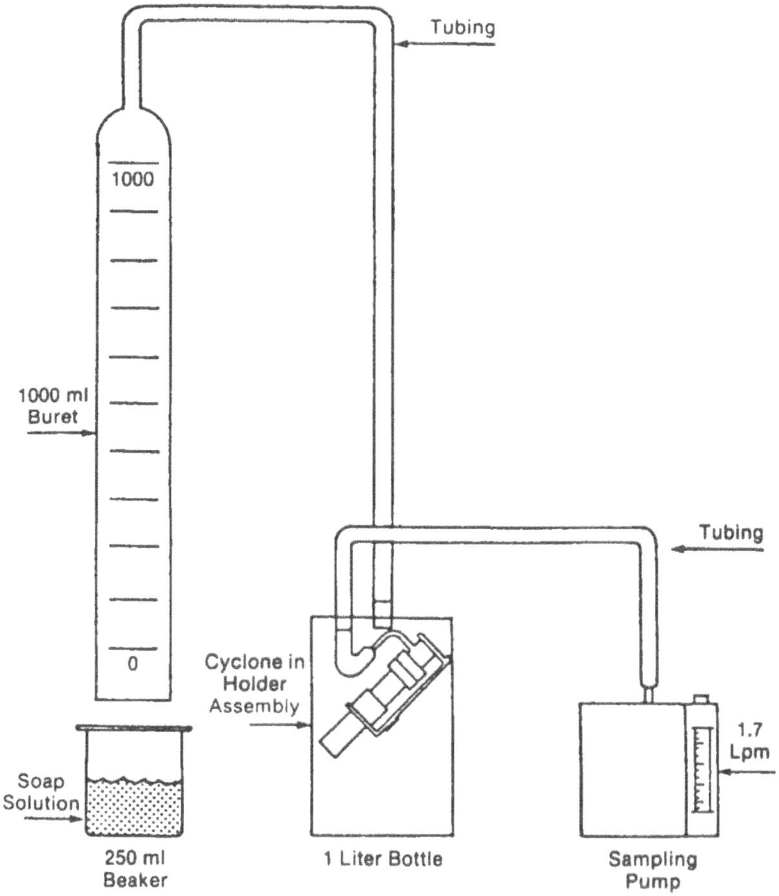

Figure 9.12. Calibration of respirable particulate matter cyclone. From *OSHA Industrial Hygiene Technical Manual*, Reference 35.

This process involves attaching the orifice unit in series with the Roots meter and a vacuum source. The vacuum source is run at a number of different flow rates — controlled either by adjusting the voltage to the motor or by blocking some of the flow through the entire system by means of restrictions between the orifice and Roots meter. (See Figure 9.13.) Roots meter volumes, times, and the orifice head loss values are recorded. The square root of the orifice head loss values (ΔH), adjusted for standard conditions, is plotted against flow rate, corrected to standard temperature (298 K) and pressure (1 atm), to yield Q_{std} (Equations 9.1 and 9.2).

$$Y = \sqrt{\Delta H \times \left(\frac{P_1}{P_{std}}\right) \times \left(\frac{T_{std}}{T_1}\right)} \qquad (9.1)$$

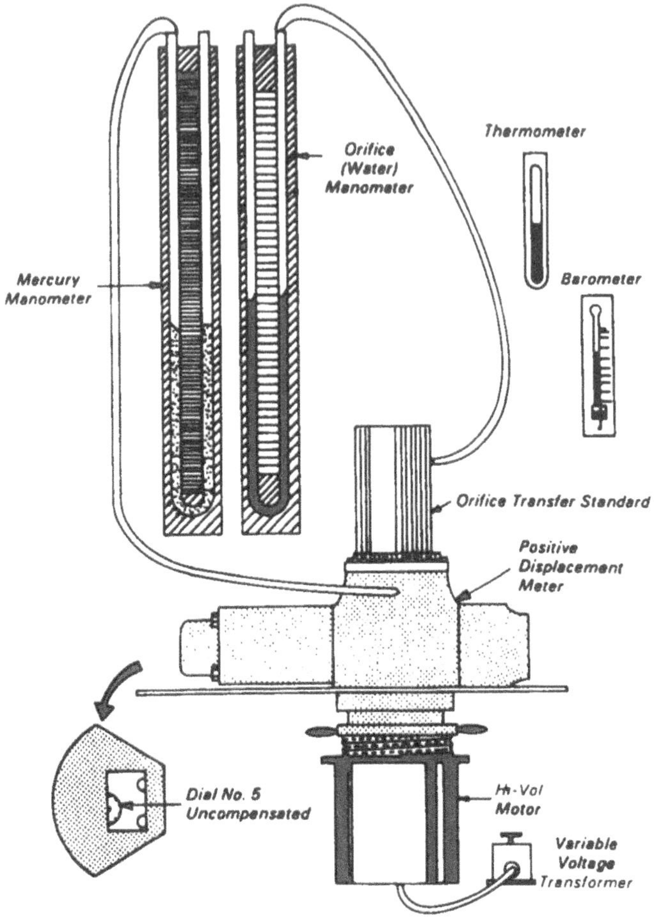

Figure 9.13. High-volume sampler calibraton, Step 1 (Reference 42).

$$Q_{std} = \frac{V_m \times \dfrac{P_1 - \Delta P}{P_{std}} \times \dfrac{T_{std}}{T_1}}{\text{time}} \quad (\text{x-axis}) \tag{9.2}$$

In these functions, T_1 and P_1 are the laboratory air temperature and pressure, respectively, and ΔP is the pressure drop or gauge pressure at the Roots meter inlet. From orifice theory, Equation 4.5b, this plot should be a straight line whose equation (Q_{std} as a function of ΔH) can be found by linear least squares regression techniques (Equation 9.3),

$$\sqrt{\Delta H \times \left(\frac{P_1}{P_{std}}\right) \times \left(\frac{T_{std}}{T_1}\right)} = m_i \times Q_{std} + b_i \tag{9.3}$$

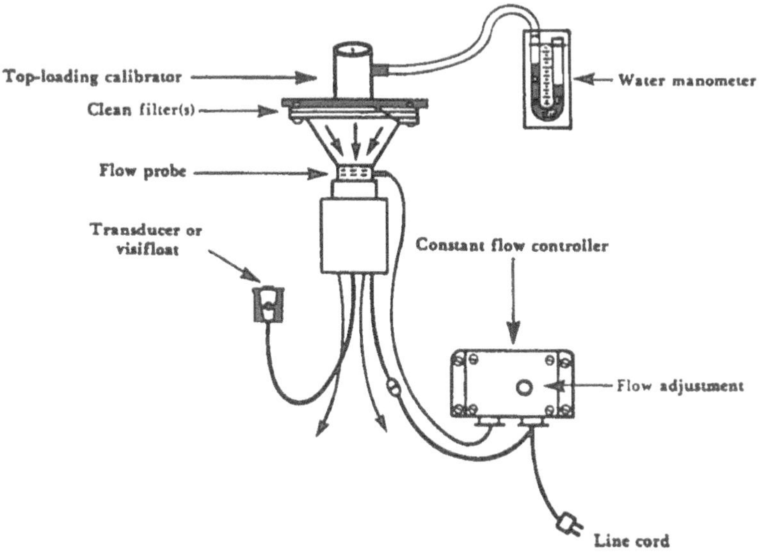

Figure 9.14. High-volume calibration, Step 2 (Reference 42).

where m_i and b_i are the slope and y-intercept, respectively. The orifice calibration equation produced should not change unless the orifice is damaged, though EPA recommends checking it once per year.

Calibration Step 2

The orifice, unlike the positive displacement meter, is a very portable device. It is used to transfer the calibration from the positive displacement meter in the laboratory to the sampler in the field, and is therefore called a "transfer calibration standard" device.

The orifice is connected to the high-volume sampler and the sampler is operated at six or more different flow rates, again controlled by voltage or flow restrictors. The data recorded in this step are the orifice head loss and the indicated sampler flow rate. The indicator of flow rate may be a rotameter or mass flow meter that samples a small portion of the total flow or a pressure transducer or another orifice meter. (See Figure 9.14.) Using the orifice head loss values and the pressure and temperature during this step, P_2 and T_2 in place of P_1 and T_1, new values for the expression in Equation 9.1 are calculated, and Equation 9.3 (solved for Q_{std}) is used to find appropriate values for Q_{std}. An electronic spreadsheet is a great help in this process.

The treatment of the values for indicated flow rate, I, depends on the type of indicator used and the method that is selected for correcting for temperature and pressure during sampling. The choices are to record actual sampling period average temperature and pressure or, if the actual data do not vary from

Table 9.1 Choices for Expressing the Indicated Flow Rate in High Volume Calibration Step 2

Indicator type	Indicator expression: Measured T_3 and P_3 (during sampling)	Indicator expression: Avg bar. pressure (P_a) and season. temperature (T_a)
Mass flow meter	I	I
Orifice/manometer	$\sqrt{I \times \left(\frac{P_2}{P_{std}}\right) \times \left(\frac{T_{std}}{T_2}\right)}$	$\sqrt{I \times \left(\frac{P_2}{P_a}\right) \times \left(\frac{T_a}{T_2}\right)}$
Rotameter or square root scale orifice and pressure recorder	$I \times \sqrt{\left(\frac{P_2}{P_{std}}\right) \times \left(\frac{T_{std}}{T_2}\right)}$	$I \times \sqrt{\left(\frac{P_2}{P_a}\right) \times \left(\frac{T_a}{T_2}\right)}$

seasonal averages by more than 60 mm Hg or 15 °C, to use the altitude-corrected annual average barometric pressure and the seasonal average temperature. Table 9.1 presents the six possible combinations.

Once it is determined which methods for flow rate recording and use of pressure and temperature data will be used, the selected adjusted flow rate indication (selected I expression) for each rate is plotted against the calculated value for Q_{std}, and again, using linear regression, the final calibration equation is obtained (Equation 9.4).

$$\text{Selected I expression} = m_{ii} \times Q_{std} + b_{ii} \tag{9.4}$$

This calibration equation is valid only for that particular combination of sampler/motor and flow rate indicator; each sampler set must be separately calibrated and will have unique values for slope, m_{ii}, and intercept, b_{ii}. Recalibration after each 6 months of use is recommended by the EPA.

Sampling

The *sampling* process requires only that the indicated flow rate be recorded (and actual pressure and temperature, if that option is chosen). For calculation of volume of air sampled, *another* table of flow rate indicator expressions (Table 9.2) and Equation 9.4, solved for Q_{std}, must be employed.

$$Q_{std} = \frac{[(\text{selected I expression}) - b_{ii}]}{m_{ii}} \tag{9.5}$$

Table 9.2 Choices for Using the Indicated Flow Rate to Compute Sampler Flow Rate

	Indicator expression	
Indicator type	**Measured T_3 and P_3 (during sampling)**	**Avg bar. pressure (P_a) and season. temperature (T_a)**
Mass flow meter	I	I
Orifice/manometer	$\sqrt{I \times \left(\frac{P_3}{P_{std}}\right) \times \left(\frac{T_{std}}{T_3}\right)}$	$\sqrt{I}$
Rotameter or square root scale orifice and pressure recorder	$I \times \sqrt{\left(\frac{P_3}{P_{std}}\right) \times \left(\frac{T_{std}}{T_3}\right)}$	I

If the sampler has a recording flow rate indicator, the average flow rate throughout the sampling period of 24 hr, Q_{std}, is calculated from the average value of I, the expression chosen from Table 9.2, and Equation 9.5. If the sampler is equipped with only an instantaneous flow rate indicator, the median of the initial and final Q_{std} values from Equation 9.5 is used. Finally, standard volume sampled is the average flow rate multiplied by sample time, and suspended particulate matter concentration for the sampling period is filter weight gain divided by volume of air sampled.

The EPA step-by-step procedure, from 40CFR50, Appendix B, is included here as Appendix F.

9.8 FLOW CALIBRATION FOR SAMPLERS WITH FLOW CONTROLLERS

There are many modifications to the procedure presented in the previous section. Flow controllers (Chapter 6) are often employed on high-volume samplers to keep the sampling rate constant. Flow control is ***required*** on PM10 and PUF samplers to ensure that the size-selecting inlet functions properly.

New flow-controlled samplers should initially be calibrated as above with the flow controller disabled. However, once a flow-controlled sampler has been shown to operate reliably, the only calibration required is a simple one-point check with an orifice transfer standard, calibrated as described in the previous section, to ensure that the flow rate is unchanged. Thus, flow controllers both produce more reliable results and reduce time involved in the calibration and sampler set-up process. Flow rate indicator values are still generally recorded during sampling operations as a quality assurance measure.

Some PM10 samplers utilizing critical orifice flow control are delivered to the user with an instrument-specific calibration flow rate table, which requires only a pressure and temperature measurement to verify that the "critical flow device" is maintaining design flow rate. Wedding[43] provides more information on critical orifice flow control in PM10 samplers. (See Figure 9.15.)

Example Problem 9.1. High-Volume Sampler Calibration

Using data given below for temperatures, pressures, orifice calibration values, and sampler calibration values, develop a calibration curve for this high-volume sampler by following the two-step calibration procedure outlined in Section 9.7 or Appendix F. Assume the actual pressure and temperature during sampling will be measured.

Orfice Calibration Data

Barometric pressure: 750 mm Hg
Laboratory room temperature: 20 °C

Orifice ΔH (" H_2O)	Roots meter volume (m^3)	Roots meter inlet gauge press (mm Hg)	Time (min)
3	2.83	−16	2.64
4	2.83	−19	2.26
5	2.83	−24	2.03
6	2.83	−36	1.86
11	2.83	−48	1.47
14	2.83	−63	1.35

Sampler Calibration Data

Barometric pressure: 740 mm Hg
Outside temperature: 30 °C

Orifice ΔH (" H_2O)	Indicator (rotameter)
16.0	49
13	45
11.0	43
7.0	34
4.4	24

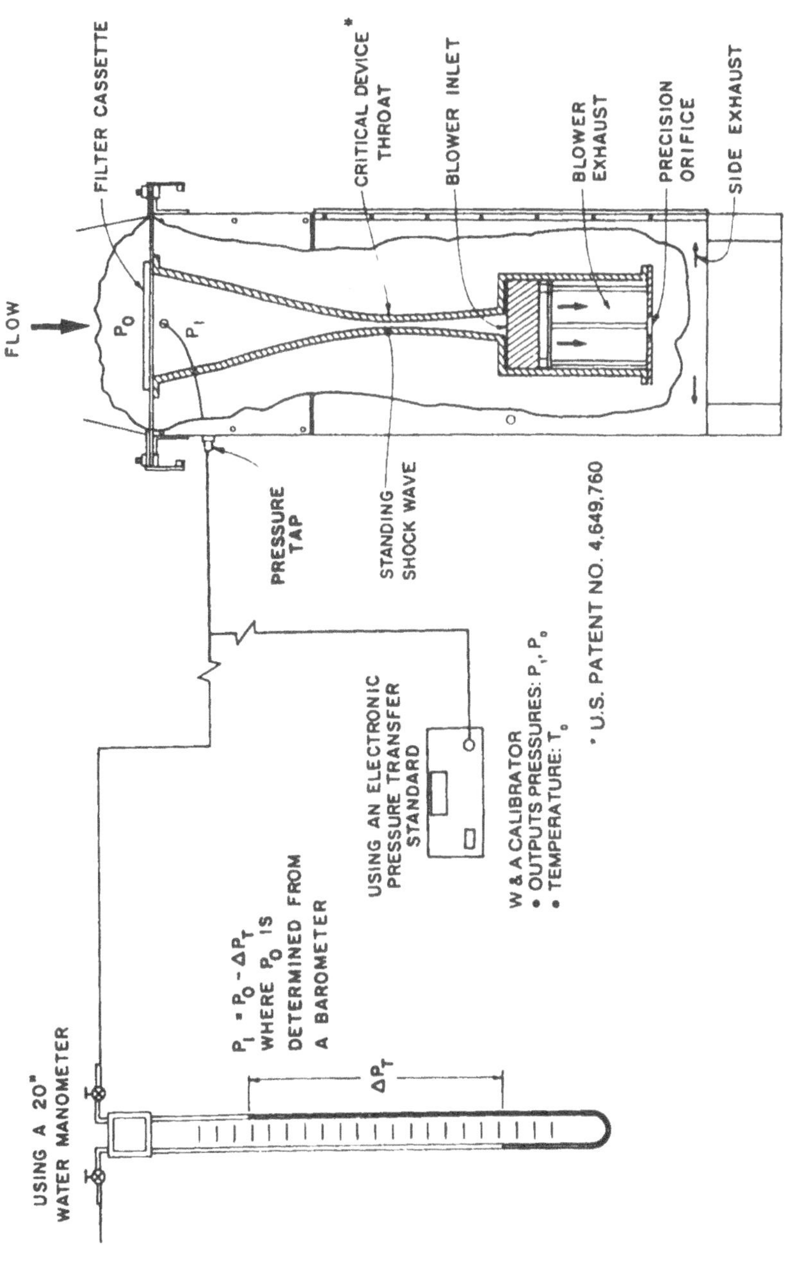

Figure 9.15. Wedding PM10 sampler. (Courtesy Wedding & Associates.)

Solution

Refer to Figure 1 in Appendix F for format. First, the orifice calibration Roots meter standard flow rates are computed using Equation 9.2. The orifice head loss values are corrected to standard conditions and raised to the $^1/_2$ power using Equation 9.1. These two pairs of values are then plotted (Q_{std} on the x-axis; corrected head loss on the y-axis), and the equation of the best fit-line found as in Equation 9.3. This can easily be done with a spreadsheet program; Figure 9.16 shows the calculated x (Q_{std}) and y (corrected head loss) values, the slope, m_i, in Equation 9.3, (labeled "x coefficient" in Figure 9.16), and the intercept, b_i (labeled "constant"). Thus, the orifice calibration equation has the form:

$$\sqrt{\Delta H \times \left(\frac{P_1}{P_{std}}\right) \times \left(\frac{T_{std}}{T_1}\right)} = 2.312 \times Q_{std} - 0.8147 \qquad \textbf{(9.6)}$$

where P_1 and T_1 are the conditions during calibration, and T_{std}, as always for ambient sampling, is 298 K. Figure 9.17 is a spreadsheet-produced plot of the data and best-fit line. For Step 2, the sampler calibration, refer to Table 2 and Figure 2 in Appendix F for format. Equation 9.1 is used with the sampler calibration pressure and temperature (P_2 and T_2) to correct the sampler calibration orifice head loss (ΔH) values. These values are successively each put in as the left-hand side of Equation 9.6 to generate a set of Q_{std} values for the sampler. Next, the corresponding indicator (rotameter) readings given are corrected as in Table 9.1. Since here the indicator is a rotameter and it is assumed that actual temperature and pressure will be measured during sampling, the indicator correction function to be selected is:

$$I \times \sqrt{\left(\frac{P_2}{P_{std}}\right) \times \left(\frac{T_{std}}{T_2}\right)} = I \times 0.9786$$

These values are plotted (y axis) against the sampler flow rate values and the best-fit line for this second step found as in Equation 9.4. This can also easily be done with a spreadsheet program; Figure 9.18 shows both the calculated Q_{std} values and corrected indicator values and the slope, m_{ii} in Equation 9.4 (labeled "x coefficient" in Figure 9.18), and the intercept, b_{ii} (labeled "constant"). The sampler calibration equation is now complete and has the form:

$$I \times \sqrt{\left(\frac{P_3}{P_{std}}\right) \times \left(\frac{T_{std}}{T_3}\right)} = 30.1212 \times Q_{std} - 12.4038 \qquad \textbf{(9.7)}$$

HIVOL SAMPLER CALIBRATION DATA: ROTAMETER

P2(mmHg)=	740	#pair cal data=	5
T2(C)=	30	303.00 (K)	

Delta H orifice	Rota Rdg (I)	Sqrt(delH* P/P*T/T)	(X)m3/min Qstd from Orif cal	(Y) I*Sqrt (P/P*T/T)	best-fit y
16.0	49.0	3.9143	2.0458	47.9504	49.219
13.0	45.0	3.5283	1.8788	44.0361	44.189
11.0	43.0	3.2456	1.7565	42.0789	40.505
7.0	34.0	2.5891	1.4725	33.2717	31.950
4.4	24.0	2.0527	1.2405	23.4859	24.961

Form of hivol motor equation:
I*sqrt(p/p*T/T)=slope*Qstd + y-int

Regression Output:

Constant	-12.4038	y-int
Std Err of Y Est	1.636208	
R Squared	0.979137	
No. of Observations	5	
Degrees of Freedom	3	
X Coefficient(s) 30.12115		<——slope
Std Err of Coef. 2.538514		

Figure 9.16. Orifice calibration results example.

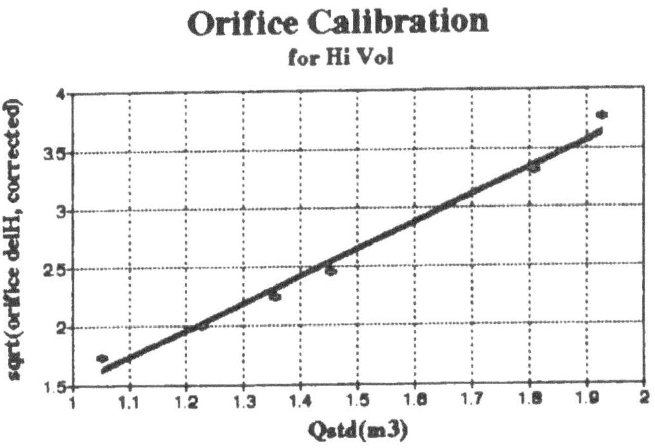

Figure 9.17. Orifice calibraton plot example.

Where T_3 and P_3 are the average temperature and pressure, respectively, *measured* during the time the high-volume sampler is collecting the sample. Figure 9.19 is a spreadsheet-produced plot of the data and best-fit sampler flow rate calibration line.

Example Problem 9.2. Calculating total suspended particulate matter sample volume and concentration

For the sampler calibrated in the previous example, compute the volume sampled and the concentration of particulate matter in the air given:

Average temperature: 2 °C
Average pressure: 737 mm Hg
Initial rotameter reading: 43
Final rotameter reading: 41
Total sample time: 1446 min
Filter mass gain during sampling: 670 mg

Solution

The initial and final sampler flow rates are found, by inputing the rotameter readings into Equation 9.7, to be 1.875 and 1.807 scmm, respectively, giving an average Q_{std} of 1.841 scmm. Dividing the mass collected by the product of Q_{std} and sample time gives an average particulate matter concentration of 251.7 $\mu g/m^3$.

HIGH VOLUME SAMPLER CALIBRATION

ORIFICE CALIBRATION DATA:

P1= 750 mmHg

T1= 20 C or in K= 293

Delta H orifice	Vroots (m3)	Proots,gage (mmHg)	Time (min)	Y sqrt(dH*P1/Pstd* Tstd/T1)	fm	X Qstd Roots	best-fit y-expect
3	2.83	16	2.64	1.735		1.053	1.619
4	2.83	19	2.26	2.004		1.225	2.017
5	2.83	24	2.03	2.240		1.354	2.316
6	2.83	36	1.86	2.454		1.454	2.546
11	2.83	48	1.47	3.323		1.809	3.366
14	2.83	63	1.35	3.749		1.927	3.640

Form of orifice equation:
[Sqrt(delH*P/P*T/T)=m*Qstd+b

Regression Output:		
Constant	-0.81474	y-int
Std Err of Y Est	0.101756	
R Squared	0.986614	
No. of Observations	6	
Degrees of Freedom	4	
X Coefficient(s)	2.311568	<——slope
Std Err of Coef.	0.134624	

9.18. High-volume calibration data example.

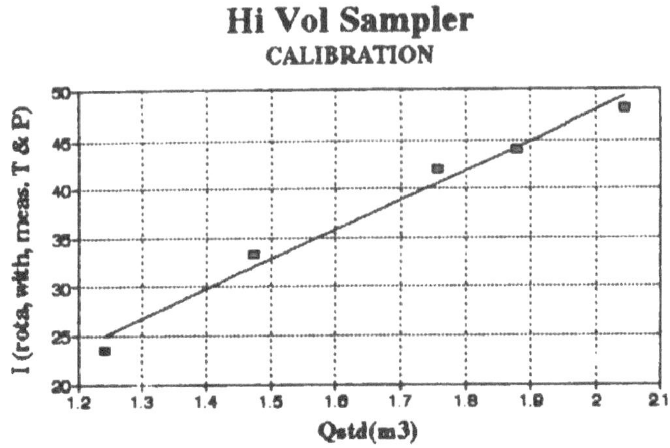

Figure 9.19. High-volume calibration plot example.

CHAPTER 9 PROBLEMS

1. Repeat Examples 9.1 and 9.2, using the same data, but with the difference that the high-volume flow rate indicator is an orifice and pressure gauge, and seasonal average temperature and altitude-corrected barometric pressure for the sampling site will be used. The indicator reading in the example will be orifice head readings in this case. For these pressure and temperature values, use P_a = 725 mm Hg and T_a = 19 °C. Use a spreadsheet program for all calculations and graphs.
2. A hi-vol calibration orifice has the linear relationship:

$$Q_{std} = \frac{\sqrt{\Delta H \times \left(\frac{P_2}{P_{std}}\right) \times \left(\frac{T_{std}}{T_2}\right)}}{m} - b$$

Given P_2 = 29.44 in Hg, T_2 = 22 °C, m = 0.859, b = 0.0266
Q_{std} = flow rate in std m³/min, ΔH = orifice head in inches H_2O,

a. What is the proper ΔH for 40 scfm?
   If the weight gain of a PM10 filter is 261 mg in 23.78 hr of sampling at 40 scfm,
b. What is the measured PM10 concentration?

3. A hi-vol calibration orifice is to be used to check a high-volume motor's flow rate. The orifice curve of Q_{std} (x-axis) vs $(\Delta H)^{1/2}$ (y-axis) is linear, with the appropriate pressure and temperature corrections employed. The slope of the plot is 1.5, and the y-intercept = 0.0266:
   a. If the orifice, P_{cal} = 29.44" Hg, T_{cal} = 22 °C, Q_{std} = flow rate in std m^3/min, ΔH = orifice head in inches H_2O, and the orifice reads 15" H_2O, what is the flow rate in std m^3?
   b. If the weight gain of a filter used with this high-vol running steadily at this flow rate is 562 mg in 24 hr of sampling, what is the measured particulate matter concentration?

10 APPLICATIONS TO SOURCE SAMPLING

10.1 INTRODUCTION

Techniques and instruments discussed in Chapters 2 through 6 are integral to the procedures known variously as "source sampling", "emission monitoring", or "stack sampling". The Code of Federal Regulations, in section 40CFR60, Appendix A,[2] spells out the EPA requirements, formally known as "EPA Reference Methods for Stationary Source Air Emissions Testing", in great detail. This chapter relates the concepts and computational techniques developed in previous chapters to the specific, unique requirements for EPA (and state regulatory agency) approved source sampling for air pollutant emissions. EPA Reference Methods 1 through 4 are basic to source sampling for most air pollutants.[6,12,14,16] The higher numbered methods, and some newer techniques that have been promulgated under incinerator emission monitoring requirements (e.g., EPA Method SW846[1]), incorporate these four basic methods by reference and go into great detail to deal with specific requirements for sampling of various pollutants, such as particulate matter, sulfur dioxide, volatile organic compounds, oxides of nitrogen, dioxins, and a whole host of other compounds.

10.2 EPA METHOD 1

Velocity measurements made with a standard or S-type pitot as described in Section 5.2, or with an electronic velocity (mass-flow) meter (Section 5.3), must be made at a number of positions in a cross-sectional plane perpendicular to the gas flow direction in a duct to fully depict the flow. EPA Method 1 (40CFR60 Appendix A)[2] gives specific guidance in selecting the proper locations in the duct to measure velocity (and, as will be discussed in Chapter 12, to sample for pollutants). The minimum number of locations needed to make measurements depends on the extent of disturbance or turbulence in the gas flow. A small number of pitot measurements will adequately characterize the gas flow if there are no "flow disturbances" in the stack or duct near the measurement site. A disturbance to flow is said to be "near the site", according to EPA Method 1, if the measurement location is less than eight times the duct diameter downstream of a bend or cross-section change that might disturb the flow lines, or less than two diameters upstream. Whichever of the two criteria is the more restrictive, eight diameters downstream or two upstream, controls the decision. The duct diameter referred to is obvious in the case of circular ducts, but it means that the "equivalent" diameter, D_e, for rectangular ducts is given by Equation 10.1,

$$D_e = \frac{2 \times L \times W}{(L + W)} \tag{10.1}$$

where L and W are the two rectangular dimensions of the duct cross-sectional area. For applications where it is not possible to meet the criteria of locating the sampling ports eight diameters downstream and two upstream of disturbances, the EPA method provides a procedure for calculating and locating a larger number of measurement locations needed to properly characterize the disturbed flow.

The EPA Method 1 procedure relies on the concept of dividing the cross-sectional area of the duct into equal areas and locating a sampling location or traverse point at the centroid of each sub-area. The minimum number of measurement points for most ducts is 12 (for small diameter ducts, as few as 8 may be used, as will be discussed below). Figure 10.1 shows the division of circular and rectangular cross sections into 12 equal areas, and the centroid location of each. Simple geometry can be used to locate the centroids; but generally, a table such as that included in 40CFR60, Appendix A, Method 1 (Table 10.1) is used to obtain distances from the inner wall in units of percent of total diameter for circular ducts, or fraction of width for rectangular ducts. If there is more variation in velocity across the cross section (i.e., it is closer to a possible flow-disturbing feature), 16 traverse points may be needed to obtain a reasonably accurate average velocity. In a manner like that depicted in Figure 10.1, the cross-section would be divided into 16 equal areas. Furthermore, in some applications, velocity in fairly small ducts may be adequately characterized by eight (for circular) or nine (for rectangular) points and, again, the concept of Figure 10.1 may be used.

To determine the appropriate number of points to use, a figure in the Code of Federal Regulations,[2] presented here as Figure 10.2, is utilized. As the figure indicates, sixteen points (8 on each of two perpendicular diameters or

LOCATION OF TRAVERSE POINTS IN CIRCULAR STACKS

[Percent of stack diameter from inside wall to traverse point]

Traverse point number on a diameter	Number of traverse points on a diameter—											
	2	4	6	8	10	12	14	16	18	20	22	24
1	14.6	6.7	4.4	3.2	2.6	2.1	1.8	1.6	1.4	1.3	1.1	1.1
2	85.4	25.0	14.6	10.5	8.2	6.7	5.7	4.9	4.4	3.9	3.5	3.2
3		75.0	29.6	19.4	14.6	11.8	9.9	8.5	7.5	6.7	6.0	5.5
4		93.3	70.4	32.3	22.6	17.7	14.6	12.5	10.9	9.7	8.7	7.9
5			85.4	67.7	34.2	25.0	20.1	16.9	14.6	12.9	11.6	10.5
6			95.6	80.6	65.8	35.6	26.9	22.0	18.8	16.5	14.6	13.2
7				89.5	77.4	64.4	36.6	28.3	23.6	20.4	18.0	16.1
8				96.8	85.4	75.0	63.4	37.5	29.6	25.0	21.8	19.4
9					91.8	82.3	73.1	62.5	38.2	30.6	26.2	23.0
10					97.4	88.2	79.9	71.7	61.8	38.8	31.5	27.2
11						93.3	85.4	78.0	70.4	61.2	39.3	32.3
12						97.9	90.1	83.1	76.4	69.4	60.7	39.8
13							94.3	87.5	81.2	75.0	68.5	60.2
14							98.2	91.5	85.4	79.6	73.8	67.7
15								95.1	89.1	83.5	78.2	72.8
16								98.4	92.5	87.1	82.0	77.0
17									95.6	90.3	85.4	80.6
18									98.6	93.3	88.4	83.9
19										96.1	91.3	86.8
20										98.7	94.0	89.5
21											96.5	92.1
22											98.9	94.5
23												96.8
24												98.9

TRAVERSE POINT	DISTANCE, % of diameter
1	4.4
2	14.6
3	29.6
4	70.4
5	85.4
6	95.6

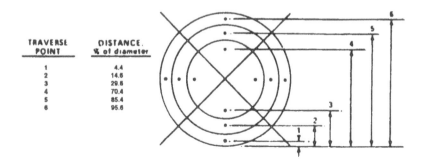

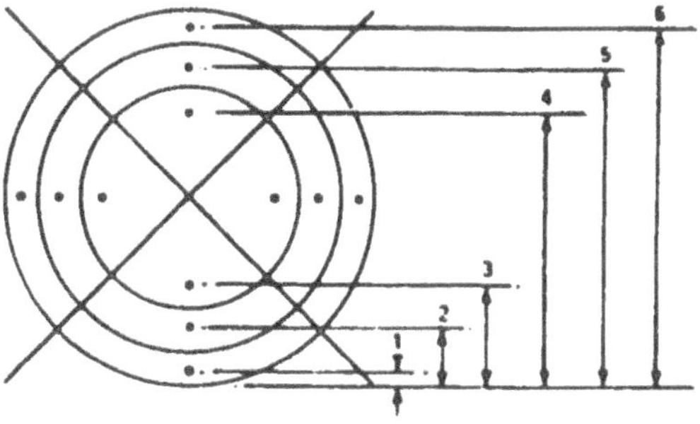

CROSS-SECTION LAYOUT FOR RECTANGULAR STACKS

Number of traverse points	Matrix layout
9	3x3
12	4x3
16	4x4
20	5x4
25	5x5
30	6x5
36	6x6
42	7x6
49	7x7

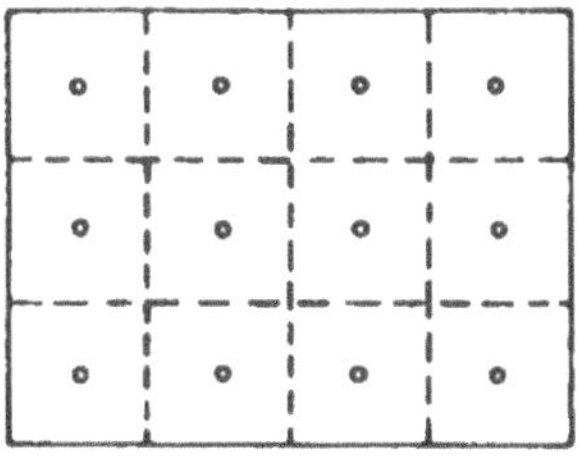

Figure 10.1. Location of traverse points (Reference 12).

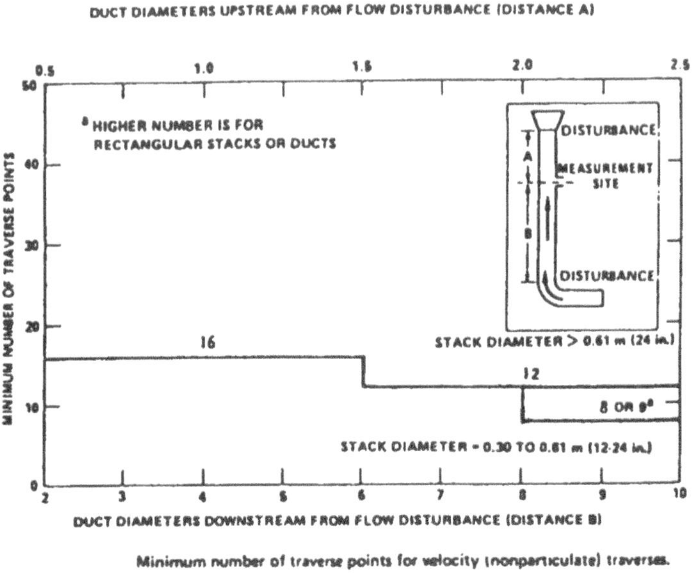

Figure 10.2. Minimum number of velocity points (Reference 12).

a 4 × 4 matrix, for circular and rectangular, respectively) are required if the sampling location is closer than 6 equivalent diameters downstream from a disturbance or 1.5 diameters upstream. The figure also shows that, for ducts with equivalent diameters under 0.61 m (24 in), fewer points may be used only at sampling location distances beyond eight diameters down or two upstream.

For velocity measurement in circular ducts, therefore, only 8, 12, or 16 points may be used, with 4, 6, or 8 on a diameter, respectively. For sampling particulate matter, many more combinations are possible. Again, Table 10.1 (from Reference 2) presents the centroidal locations as percent of diameter distances from the duct inner wall for many possible combinations from 8 (4 on each diameter) to 32 (16 each diameter). Figure 10.3 presents the more complicated criteria for number of points when sampling for particulate matter.

An important aspect of Method 1 is the requirement that any traverse point locations, as calculated from Figures 10.2 or 10.3 and Table 10.1, that fall within 2.5 cm (1") from the duct wall (or 1.3 cm for ducts with diameters less than 0.61 m) must be moved to 2.5 cm from the wall (or again, 1.3 cm for small ducts) and called "adjusted" traverse points.

Example Problem 10.1. EPA Method 1 Calculation

A horizontal circular duct with an inner diameter of exactly 1.2 m is to be evaluated for velocity and flow rate. There is a sampling location which

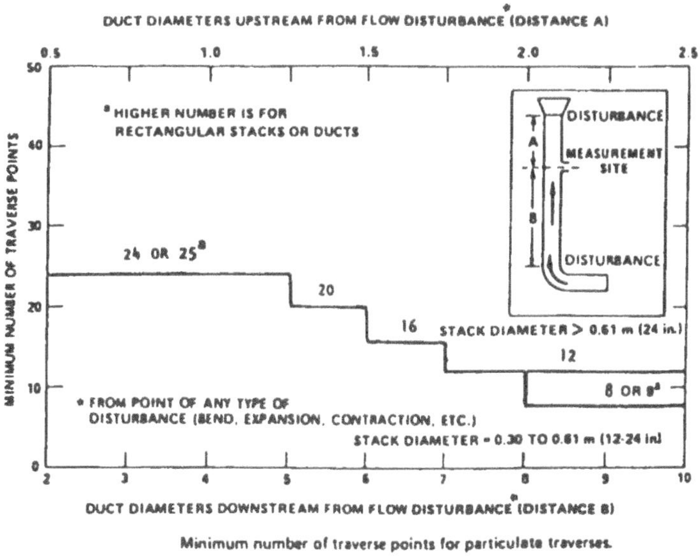

Figure 10.3. Minimum number of sampling points (Reference 12).

is 9.8 m downstream from a bend in the ductwork and also 1.7 m from where the duct enters the vertical stack. Locate the velocity measurement points.

Solution

Figure 10.2 indicates for 9.8/1.2 = 8.2 diameters downstream: 12 points; but for 1.7/1.2 = 1.4 diameters upstream: 16 traverse points. Thus, 16 points must be used. Table 10.1 shows that these 16 points will be located, 8 on each of two perpendicular diameters, at:

1. 3.2% of 1.2 m = 0.038 m or 3.8 cm from the wall.
2. 10.5% 12.6 cm
3. 19.4 23.3 cm
4. 32.3 38.8 cm
5. 67.7 81.2 cm
6. 80.6 96.7 cm
7. 89.5 107.4 cm
8. 96.8 116.2 cm

Example Problem 10.2. EPA Method 1 for Particulate Sampling

Repeat the previous Example Problem for particulate sampling in the same duct.

Solution

As in the previous Example Problem, the sampling location is 8.2 diameters downstream, and 1.4 diameters upstream of a disturbance. For particulate sampling, Figure 10.3 must be used — indicating 12 points for the downstream criterion, but 20 points for the upstream criterion. Table 10.1 gives the percentage of diameter for each point in the column under 10 points on a diameter.

For very small ducts (0.071 to 0.30 m^2 or 4 to 12 in^2 cross-section), some modifications to Method 1 are required due to the potential for the normal sampling probe to significantly block and disturb the exhaust gas flow. A standard pitot is employed for velocity measurement at a location downstream from the normal sampling point. Refer to 40CFR60, Appendix A,[2] Method 1A, for details.

EPA Method 1 also presents alternate techniques for another special situation: ducts which exhibit swirling or cyclonic flow. The 40CFR60, Appendix A, should be consulted for this issue.

10.3 EPA METHOD 2

Velocity measurements with a standard or a Stausscheibe pitot tube as discussed in Section 5.2 are applied in EPA Emission Monitoring Method 2. A standard pitot is used for calibration and for very small ducts, while an S-type pitot is used for most emission sampling.

As noted in Section 5.2, each leg of an S-type pitot will have its own unique coefficient, $C_{p(s)}$. For use of an S-type pitot without regard to which leg is facing upstream, the two coefficients (one for each orientation of the symmetrical S-type pitot) must be nearly equal. If the pitot is constructed so as to be precisely symmetrical, the two coefficients should both be close to 0.84. However, S-type pitot tubes are often rigidly mounted on a stack sampling probe, which may cause flow disturbance around the pitot such that the coefficient differs from 0.84. Figures 10.4 and 10.5, from 40CFR60 Appendix A, Method 2,[2] show the proper construction of a probe-mounted pitot tube. Any probe-mounted pitot tube should have coefficients for both orientations found by using a calibration set-up as shown in Figure 5.6 and calculated from Equation 10.2.

$$C_{p(s)} = 0.99 \times \left(\frac{\Delta P_{std}}{\Delta P_s} \right)^{\frac{1}{2}} \tag{10.2}$$

Typically, three sets of calibration data, at three different gas flow rates, are averaged to compute the average coefficient. If the legs are labeled A and B, a requirement of Method 2 is that the average coefficients must be within 0.01 units; that is,

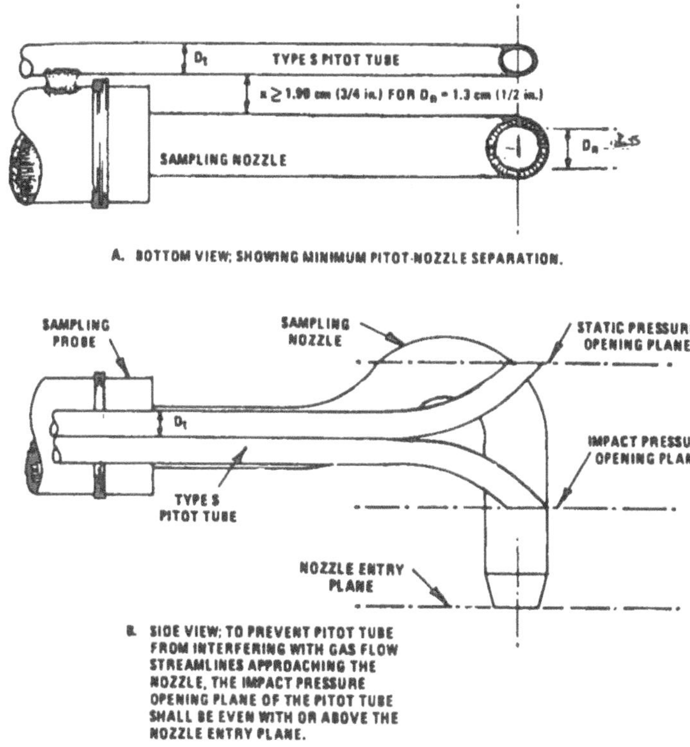

Figure 10.4. Proper probe-pitot configuration (Reference 12).

$$\left| C_{p,A,avg} - C_{p,B,avg} \right| \leq 0.01 \tag{10.3}$$

Furthermore, the calibration data for each leg must show little scatter, as measured by the average deviation, σ, the average of the difference between the average coefficient and each of the individually computed coefficients for that orientation:

$$\sigma = \frac{\left[\Sigma \left| C_{p(i)} - C_{p,avg} \right| \right]}{n} \tag{10.4}$$

where n is the number of sets of calibration data for that orientation (usually 3), $C_{p(i)}$ is an individual value from Equation 10.2, and $C_{p,avg}$ is the average of all n coefficients. The values for σ for both orientations are calculated separately, and both must be less than 0.01:

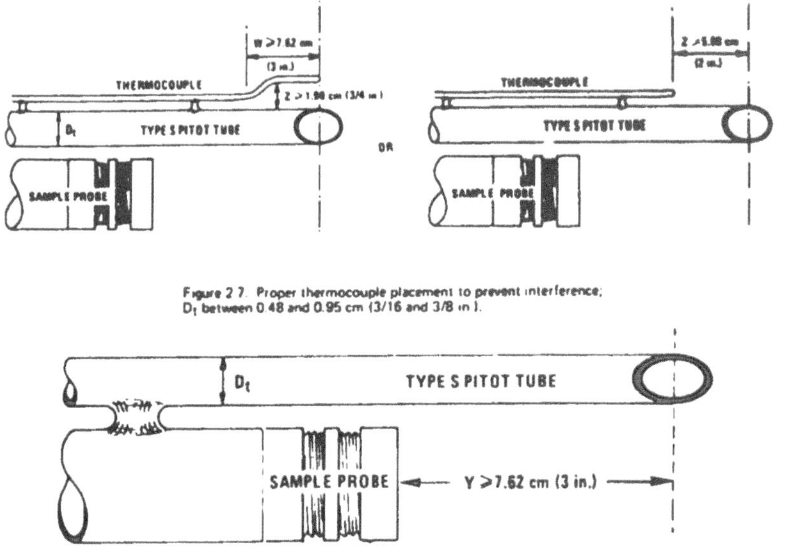

Figure 10.5. Interference free pitot and probe (Reference 12).

$$\sigma \text{ (both A and B)} \leq 0.01 \tag{10.5}$$

If the pitot does not comply with Equations 10.3 and 10.5, the pitot must have its configuration checked and adjusted and be recalibrated.

Example Problem 10.3. Evaluation of acceptability of pitot calibration

Compute the average deviation and the difference between the two S-type pitot coefficients for the pitot calibration data of Example Problem 1 in Section 5.2.

Solution

The computed coefficients were:

Leg A upstream	0.857	
	0.844	
	0.839	Average = 0.847
Reversed (B)	0.848	
	0.824	
	0.832	Average = 0.835

The difference between the two average coefficients is 0.012, just a bit too large to meet the Method 2 criterion.

The deviations are:

Leg A $|0.847 - 0.857| = 0.010$
$|0.847 - 0.844| = 0.003$
$|0.847 - 0.839| = 0.008$

Average deviation (A): $\sigma(A) = 0.007$

Leg B $|0.835 - 0.848| = 0.013$
$|0.835 - 0.824| = 0.011$
$|0.835 - 0.832| = 0.003$

Average deviation (B): $\sigma(B) = 0.009$

Both deviations are less than 0.01, so the calibration of this pitot meets one criterion, but fails the requirement that the experimental coefficients for both directions be within 0.01 units of each other.

Once the calibration is complete and acceptable $C_{p(s)}$ documented, the pitot may be used to make velocity measurements at locations in the duct cross-section specified by Method 1, and Equation 5.7b may be used to calculate fluid velocity at each location in the flowing gas stream where the pitot measurement is made.

$$v_s = C_p \times K_p \times \left[\frac{\left(T_{gas} \times \Delta P_p\right)}{P_{stat} \times M_{gas}}\right]^{\frac{1}{2}} \tag{5.7b}$$

The absolute pressure of the gas, P_{stat}, is generally uniform at all traverse points, so a single P_{gage} measurement and barometric reading suffice; but the temperature, T_{gas}, must be measured at each point with a thermocouple attached to the pitot. The molecular weight of the gas must also be known; EPA Method 3 provides a measurement technique. Depending on the set of units chosen, velocity of the stack gas will typically be in meters per second (m/sec) or feet per second (ft/sec).

The volume flow rate of the gas may also be computed, typically in units of dry standard cubic meters or feet per hour (dscmh or dscfh, respectively). Information needed to complete this calculation includes the average gas temperature, absolute pressure, the duct cross-sectional area, and the moisture content of the gas, B_{H_2O}, (measured by Method 4).

$$Q_{s(dry,std)} = 3600 \times \left(1 - B_{H_2O}\right) \times v_s \times A \times \left(\frac{P_{stat}}{P_{std}}\right) \times \left(\frac{T_{std}}{T_{s)avg)}}\right) \tag{10.6}$$

where
$Q_{s(dry,std)}$ = flow rate in dry, standard m^3/hr or ft^3/hr
B_{H_2O} = water vapor volume fraction of gas
T_{std} = 293 K or 528 °R for emission sampling
A = cross-sectional area, m^2 or ft^2

10.4 EPA METHOD 3

In the previous section, the calculation of gas velocity in the duct or stack with a pitot tube required measurement of the pitot velocity head, ΔP, and several other parameters, (including the molecular weight of the gas) for insertion into Equation 5.7b:

$$v_s = C_p \times K_p \times \left[\frac{\left(T_{gas} \times \Delta P_p\right)}{\left(P_{stat} \times M_{gas}\right)}\right]^{\frac{1}{2}}$$

EPA Method 3 addresses techniques to measure gas properties that allow computation of the average stack gas mixture molecular weight for use in the velocity calculation, as well as other attributes of the gas.

The primary constituents of exhaust gas from most combustion or ventilation systems are nitrogen, carbon dioxide, oxygen, carbon monoxide, and water vapor. The pollutants to be measured are mixed into this "carrier gas" mixture in much smaller quantities than any of the primary constituents. There are a number of devices which can measure the volume percentages of the major constituents to allow computation of the average dry molecular weight of the mixture. The device that has been most used for this purpose over the past 30 years is the Orsat analyzer shown in Figure 10.6.

Orsat Analyzer

The Orsat device is an apparatus that allows a known volume sample of gas to be trapped over water (the sample is saturated with water) and then successively transferred to three chambers that contain solutions of potassium hydroxide (absorbs CO_2), pyrogallic acid (absorbs O_2), and cuprous chloride (absorbs CO).[4] After each absorption step, the volume of gas remaining is measured, and the partial volume of each gas absorbed, divided by the total original volume (this quotient is called the volume fraction, denoted B_i for constituent i), is recorded. It can easily be shown, through Ideal Gas Law relationships, that the volume fraction of a gas in a mixture is equal to both its mole fraction and its partial pressure ratio. The absorption of one constituent of the gas in each one of the absorption chambers reduces the pressure of the remaining mixture by that gas's partial pressure. The use of water to capture and transfer the gas in the device ensures constant saturation, and also constant temperature, and the return of the sample gas to its original pressure before each measurement of quantity absorbed ensures accurate volume fraction measurements. The device, therefore, reports volume or mole fractions of three of the major constituents on a "dry basis" (water ignored) and allows water vapor to be deleted from the computation.[23]

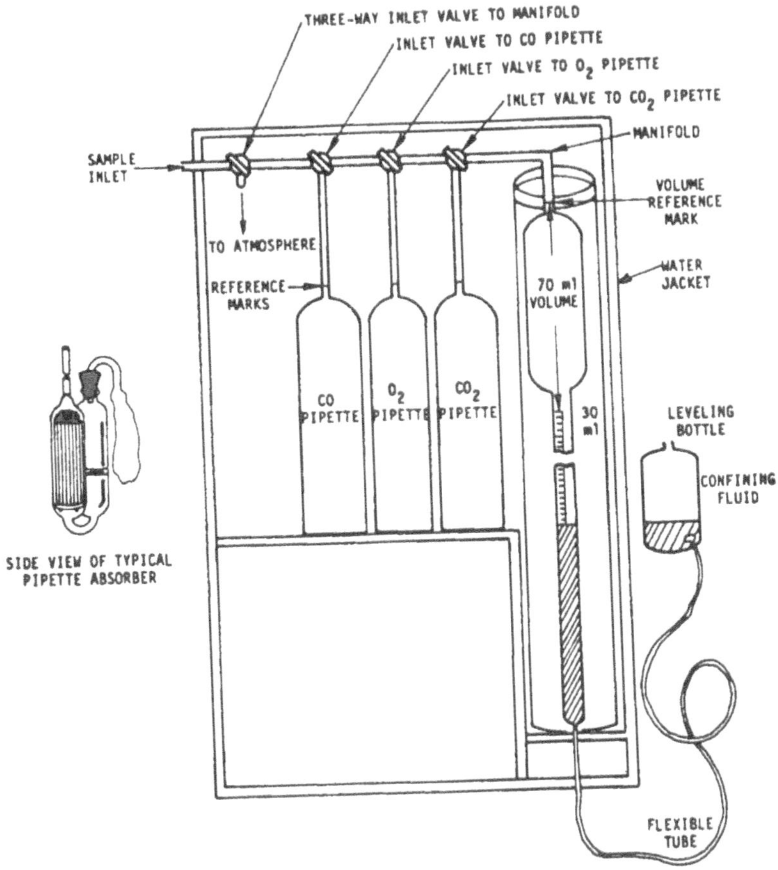

Figure 10.6. Orsat analyzer.

Since the only remaining major constituent of the gas, in most instances, is nitrogen (N_2), it is assumed that all the remaining gas is N_2, and the data resulting from the Orsat measurement are four dry basis volume fractions: B_{CO_2}, B_{O_2}, B_{CO}, and B_{N_2}.

While the chemical analysis of the Orsat apparatus is precise if the device is properly maintained and operated, it is a cumbersome and time-consuming apparatus to use. There are several popular alternative methods for making the measurements needed to compute the dry molecular weight of the exhaust gas.

Alternatives to Orsat

One easy to use adaptation of the Orsat technique is the Fyrite flue gas kit manufactured by the Bacharach Corporation (Figure 10.7). EPA Method 3

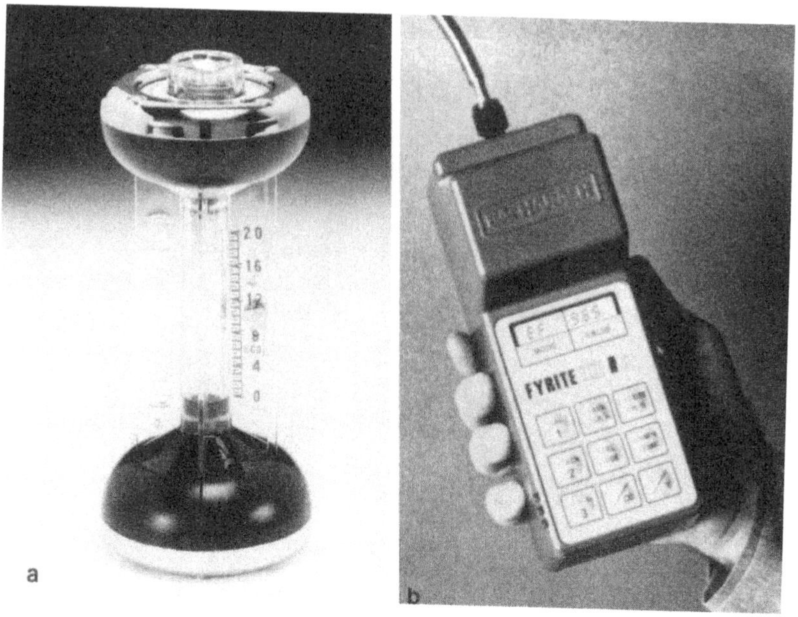

Figure 10.7. Fyrite flue gas analyzer (a) and Fyrite II (b). (Courtesy Bacharach, Inc., Pittsburgh, Pennsylvania.)

allows the use of this type of device for dry molecular weight determination. Since in most situations carbon monoxide concentration is at levels below a few hundred parts per million (1% = 10^4 ppm) and below the detection limit of an Orsat, the Fyrite kit is used to measure only O_2 and CO — a separate absorption vessel for each gas is included in the kit. The volume fraction of CO is generally negligible. Techniques and results are comparable to those of Orsat.

Electric or battery-powered gas analyzers use various techniques to detect CO_2, O_2, and CO, and are specifically authorized for air pollutant emission work in EPA Reference Method 3A.[2] Several manufacturers produce devices that measure CO_2 and CO by taking advantage of their absorption of infrared light at unique wavelengths. Automotive exhaust gas inspection devices and continuous emission monitors employ this technique and are very adaptable to EPA Method 3 stationary source emission measurements. O_2 can be measured by means of its solubility in aqueous solutions or a number of other techniques adaptable to compact electronic devices. Instruments designed for continuous monitoring of emissions of CO_2, O_2, and CO employ the same principles and are often employed in lieu of the Orsat.

Dry Molecular Weight Calculations

The four volume fractions (B_i values) found from Orsat analysis or alternative methods are equal to their mole fractions. For example, $B_{O_2} = N_{O_2}/n_{mix}$, where

N_{O_2} = number of moles of oxygen in the gas and n_{mix} = total moles of mixture. Therefore, multiplying each B_i value by the molecular weight or mass per mole, M_i, of that constituent gives the mass of that constituent divided by the total number of moles of mixture; for example,

$$B_{O_2} \times M_{O_2} = \left(\frac{n_{O_2}}{n_{mix}}\right) \times M_{O_2} = \frac{n_{O_2} \times M_{O_2}}{n_{mix}} = \frac{m_{O_2}}{n_{mix}} \quad (10.7)$$

where m_{O_2}, the mass of oxygen in the mixture, is the product of number of moles and the mass per mole, or molecular weight. If each volume fraction is multiplied by its molecular weight, and the products added, the result will be the total mass divided by the total moles of mixture, m_{mix}/n_{mix}, which is the mixture's weighted average mass per mole or the average mixture molecular weight. Recalling that water vapor was removed from consideration in this analysis, the resulting sum is called the "dry molecular weight" of the mixture of gases, M_{dry}.

$$\sum(B_i \times M_i) = M_{dry} \quad (10.8)$$

This average molecular weight of the gas mixture is not the value needed in the pitot velocity equation, but is a necessary step in computing that value. EPA Method 4, covered in the next section, allows the completion of the total molecular weight calculation, and thus the velocity and gas flow rate computations for Method 2.

Example Problem 10.4. Dry Molecular Weight Calculation

An Orsat analysis, or one of the electronic instrument alternatives, gave the following analysis for the exhaust gas from a diesel generator engine: 11.4% CO_2, 5.6% O_2, and 450 ppm CO. Compute the dry molecular weight of the gas.

Solution

The percents given are 100 times the B_i values. The CO concentration must be multiplied by 10^{-4} to obtain 0.045%. Adding the percents and subtracting from 100 gives for nitrogen, 82.955 or about 83%. Multiplying each fraction by molecular weight, M_i, and using Equation 10.8, one obtains:

$$\begin{aligned} 11.4/100 \times 44 &= 5.02 \\ + 5.6/100 \times 32 &= 1.79 \\ + 0.045/100 \times 28 &= 0.01 \\ + 83/100 \times 28 &= 23.24 \\ \text{Dry molecular weight of gas} = \Sigma B_i \times M_i &= 30.06 \end{aligned}$$

The units of M_{dry} may be selected to conform to the user's unit system: g/gmole, kg/kgmole, or lb/lbmole. Numerically, all three have the same value. To use the molecular weight of the exhaust gas in the equation for velocity (Equation 5.7b), the moisture content of the gas must be measured, and a further application of Equation 10.8 employed to compute the "total molecular weight" of the gas, M_{gas}.

10.5 PERCENT EXCESS AIR AND AIR/FUEL RATIO CALCULATION

The data from the Orsat or alternative measurements have other applications beyond average molecular weight calculation. The data may be used to compute air-to-fuel ratios or calculate percent of excess air for combustion sources, both of which are parameters which relate to proper and efficient operation of the fuel-burning source.[20] The percent or fraction of CO_2 and/or O_2 can also be used to correct the measured concentration of pollutants in the exhaust gas to standard levels of dilution, as often required by state or federal air pollutant emission regulations. Examples of this dilution-correction type of calculation are presented in the Chapter 13, which deals with calculation of emission rate.

The percent excess air, %EA, is a measure of the amount of air mixed with a given quantity of fuel beyond that theoretically required to combust all of the fuel molecules and oxidize all the carbon, hydrogen, and other oxidizable atoms in the fuel completely. In all real combustion processes, a substantial amount of excess air is required because the fuel droplets or particles are of finite size, and each must be surrounded by more than the theoretical number of oxygen molecules to ensure oxidation of all the carbon and/or hydrogen atoms. It should be obvious that the %EA is directly related to the air-to-fuel ratio. For example, 25% excess air (25% EA) means that an additional 25% more air is supplied than the theoretical amount of air for each molecule of fuel. The actual air-to-fuel ratio is thus 25% higher than the theoretical or stoichiometric ratio. In simple situations, %EA can be calculated directly from Orsat or other exhaust gas analysis data using Equation 10.9.

$$\%EA = \left[\frac{(\%O_2 - 0.5 \times \%CO)}{(0.264 \times \%N_2 - \%O_2 + 0.5 \times \%CO)}\right] \times 100 \qquad (10.9)$$

Example Problem 10.5. Computing %EA from Orsat Data

Using the same Orsat analysis data as in Example 4, 11.4% CO_2, 5.6% O_2, and 450 ppm CO — compute the %EA.

Solution

Using Equation 10.9,

$$\%EA = \{(5.6 - 0.5 \times 0.045)/[0.264 \times 82.955 - 5.6 + 0.5 \times 0.045]\} \times 100$$

$$\%EA = 34$$

If the fuel contains oxygen (e.g., methanol or "gasohol"), Equation 10.9 will give incorrect results. Whether or not the fuel contains oxygen, the %EA can be found by using the Orsat data to balance the fuel combustion equation, computing the air-to-fuel ratio (A/F), and calculating the %EA from Equation 10.10,

$$\%EA = \left[\frac{(A/F)_{act}}{(A/F)_{stoich}} - 1\right] \times 100 \qquad (10.10)$$

where $(A/F)_{stoich}$ is the stoichiometric or theoretical air to fuel ratio and $(A/F)_{act}$ is the actual ratio of air to fuel. To find the two air-to-fuel ratios, it is necessary to balance two chemical reactions, using the Orsat data. Example Problem 6 demonstrates the technique.

Example Problem 10.6. Computing %EA

Using the same Orsat analysis data as in Example Problem 4 — 11.4% CO_2, 5.6% O_2, and 450 ppm CO — compute a) the A/F on a mass basis, and b) the %EA.

Solution

a) Begin by assuming the fuel is a pure hydrocarbon of unknown makeup, C_xH_y. The fuel is likely to be a mixture of several pure hydrocarbons, so x and y will be average values, not whole numbers. The combustion equation for this theoretical molecule is balanced for the production of 100 moles of dry exhaust gases (i.e., the Orsat percents are taken as moles of each gas). In Example 4, N_2 was found to constitute 82.955% of the dry products.

$$C_xH_y + a\,O_2 + 3.76 \times a\,N_2$$

$$= 11.4\,CO_2 + 0.045\,CO + 5.6\,O_2 + 82.955\,N_2 + b\,H_2O$$

where a is the unknown number of moles of O_2 required to produce the measured products, 3.76 is the ratio of N_2 to oxygen in air, and b is the unknown number of moles of H_2O produced.

A carbon balance gives x = 11.4 + 0.045 = 11.445 carbon atoms per molecule of this unknown fuel. A nitrogen balance reveals:

$$3.76 \times a = 82.955, \text{ or } a = 22.06.$$

An oxygen atom balance is then possible:

$$2 \times 22.06 = (2 \times 11.4) + (1 \times 0.045) + (2 \times 5.6) + b,$$

and solving for b: $b = 10.08$.

Finally, balancing the hydrogen atoms on each side yields: y = 2 × b = 20.15.

With all the coefficients in the equation known, the $(A/F)_{act}$ can be computed: moles of air per mole of fuel = a + (3.76 × a) = (4.76 × a) = (4.76 × 22.06) = 105.01. Thus, $(A/F)_{mole}$ basis = moles air/mole fuel = 105.01/1. To find the $(A/F)_{act}$ on a mass basis, the moles of each must be multiplied by the molecular weight. The fuel molecular weight is [(atomic wt carbon × 11.445) + (atomic wt hydrogen × 20.15)] = [(12 × 11.445 + 1 × 20.15)] = 157.49, and molecular weight of air is 28.96. Thus,

$$(A/F)_{act} = (105.01 \times 28.96)/(1 \times 157.49) = 19.31 \text{ kg air/kg fuel}$$

For part b), the %EA, the stoichiometric air-to-fuel ratio must first be found, by a process of balancing the theoretical combustion equation for the same fuel. The theoretical combustion results in only CO_2, N_2, and H_2O in the products.

$$C_{11.445}H_{20.15} + c\,O_2 + (3.76 \times c\,N_2) = d\,CO_2 + e\,N_2 + f\,H_2O$$

The carbon balance gives d = 11.445. Balancing the hydrogen atoms on each side yields:

$$f = 20.15/2 = 10.08$$

An oxygen atom balance is then possible:

$$2 \times c = 2 \times d + f = 2 \times 11.445 + 10.08$$

and solving for c:

$$c = 16.49$$

The nitrogen balance gives 3.76 × c = 61.98 = e
With all the coefficients in the theoretical combustion equation known, $(A/F)_{stoich}$ can be computed:

$$\text{moles of air per mole of fuel} = c + (3.76 \times c) = 78.49$$

Thus, $(A/F)_{stoich}$ on a mass basis =
(78.49 × 28.96)/(1 × 157.49) = 14.43 kg air/kg fuel.
Finally, the %EA is found from Equation 10.12:

$$\%EA = \left[\frac{(A/F)_{act}}{(A/F)_{stoich}} - 1\right] \times 100 = \{19.31/14.43 - 1\} \times 100$$

or

$$\%EA = 34\%$$

This is the same answer found in Example Problem 5.

If the fuel contains oxygen, the technique of Example Problem 6 must be used, but additional information on the oxygen content of the fuel is required. See Problem 10.3.

10.6 EPA METHOD 4

Section 10.5 discussed the use of empirical data on the makeup of exhaust gas to compute the dry molecular weight of the gas. In air pollutant emissions testing, the ultimate use of the molecular weight is in the calculation of the gas velocity and flow rate. For this purpose, however, the total or "wet" molecular weight is needed, not the dry molecular weight. It is the purpose of EPA Reference Method 4[2] to measure the gas moisture or H_2O content and allow the calculation of total molecular weight.

The EPA Reference Method for measurement of moisture content in a gas stream is a combined condensation and adsorption method. A sample of the gas is first drawn through a heated probe and/or sampling line, where its temperature is kept above the dewpoint to prevent any condensation. The gas then passes into the condenser, where its temperature is brought below the dewpoint and water vapor is allowed to condense out. The gas then passes through a hygroscopic medium (usually silica gel adsorbent), where the remaining water vapor is removed. The dry gas sample is then passed through a dry gas meter, where its temperature, pressure, and volume are measured. The collection of apparatus or "sampling train" generally used for this process is designed for convenience of operation, and is generally the same equipment used for other stack sampling methods (e.g., Methods 5, 6, 7, 10,[2] or SW846[1] methods). A

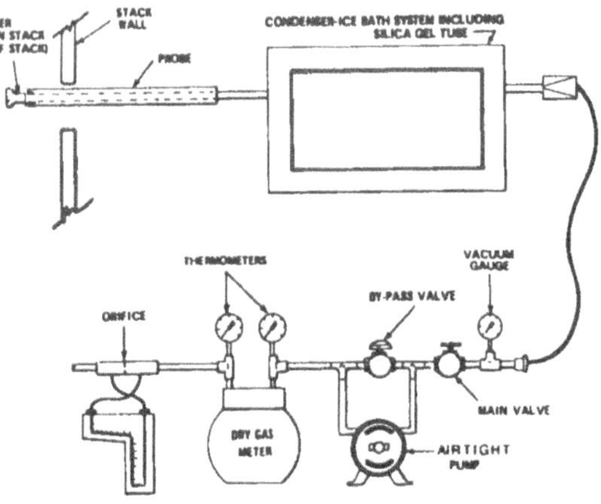

Figure 10.8. Schematic of Method 4 (Reference 12).

schematic of the sampling system is shown in Figure 10.8, and a photograph of the actual apparatus is seen in Figure 10.9.

There are a number of specific requirements for the equipment. Since the objective is to measure accurately the water vapor in the condenser/adsorber section of the apparatus, the probe and sample lines upstream of this section must be inert and heated to avoid condensation, and the whole system must be leak-free. The specific requirements are spelled out in great detail in 40CFR60, Appendix A, Method 4,[2] and will be reviewed here.

Figure 10.10 shows the condenser/adsorber box. Various manufacturers produce slightly different configurations, but the primary design requires leak-free connection to the sample line and four glass impingers connected in series and installed in an ice bath. The first two impingers are filled with an accurately measured quantity of water and act as bubblers; the gas is drawn down through the cold water and bubbles up, then travels out to the next impinger. There are specific requirements on the impinger design (the impingers are specifically known as "Greenburg-Smith" impingers or modifications) that help to control size of gas bubbles and thus the efficiency of heat transfer and condensation. The third impinger is left dry for further condensation; the fourth impinger contains a quantity of silica gel adsorbent that removes nearly all the remaining water vapor as the gas passes through before finally exiting. The dry gas continues through the pump and on to the dry gas meter, where its volume is accurately recorded.

The specific sampling requirements involve selecting the proper traverse points in the emission duct cross-section using Method 1 (see Section 10.2), assembling and leak-checking the sampling lines and impingers, and sampling at a flow rate of approximately 0.021 m^3/min (0.75 ft^3/min). Sampling should be performed for an equal amount of time and volume at each of the Method

Figure 10.9. EPA Method 4 sampling equipment.

1-identified points and for a total sample volume of at least 0.6 standard m^3 (21 standard ft^3, scf), as calculated using the dry gas meter reading, the correction factor C_{dgm} or Y (Equations 3.10b or 3.11), and Boyle's and Charles' Laws (Equation 2.12). The temperature and absolute pressure of the dry gas meter at each sampling point must be recorded. In most commercial sampling trains, the temperature and pressure information is easily obtained. Again, 40CFR60, Appendix A, gives complete details.

After sampling is complete, the apparatus is dismantled and the quantity of H_2O collected from the sampled gas is measured by the increase in the total volume of the water in the first three impingers and the increase in mass of the silica gel adsorbent. The quantities measured are the initial and final water volumes, V_i and V_f; the initial and final silica gel masses, m_i and m_f; and the volume of dry gas measured by the dry gas meter. Through the use of the Ideal Gas Law, the water vapor volume at standard temperature of 20 °C (293 K or 528 °R) the standard pressure of 1 atm (760 mmHg or 29.92" Hg), and the dry gas volume at the same conditions can be calculated as shown below. With this information, the volume fraction of water vapor in the gas B_{H_2O} can be

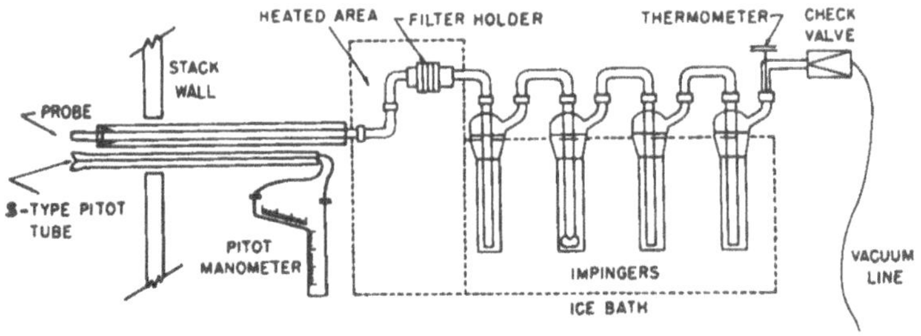

Figure 10.10. Method 4 impingers (Reference 12).

computed, and this value used to find the total molecular weight of the gas. The five basic steps are given below.

1. Volume of water vapor condensed is found through an application of Equation 2.5:

$$PV = (m/M) \times R_u T$$

where R_u = 0.06236 mm Hg-m³/gmole-K or 21.85" Hg-ft³/lbmole °R
$T = T_{std}$ = 528 °R or 293 K
P = 29.92" Hg or 760 mm Hg
M = molecular weight of water, 18 g/gmole or 18 lb/lbmole
m = mass of water condensed.

The water mass is equal to $= \rho_w \times (V_f - V_i)$, where ρ_w = density, V_f = final liquid water volume, and V_i = initial liquid water volume. Thus, the volume of water vapor at standard temperature and pressure is given by Equation 10.11,

$$V_{wc,(std)} = \frac{(V_f - V_i) \times \rho_w \times R_u \times T_{std}}{P_{std} \times M_w} = K_1 \times (V_f - V_i) \quad \textbf{(10.11)}$$

where K_1 = 0.001333 std m³/mL = 0.04707 std ft³/mL

2. Similarly, the volume of vapor adsorbed onto the silica gel is calculated from the mass increase:

$$V_{wsg,(std)} = \frac{(m_f - m_i) \times R_u \times T_{std}}{P_{std} \times M_w} = K_2 \times (m_f - m_i) \quad \textbf{(10.12)}$$

where m_f = final mass of silica gel
m_i = initial mass of silica gel
K_2 = 0.001335 std m³/g = .04715 std ft³/g

3. At the same temperature and pressure, the volume of dry gas is computed from the dry gas meter reading, its correction factor, and its average operating temperature and pressure:

$$V_{dgm,\,(std)} = V_{dgm} \times Y \times \left(\frac{P_{dgm} \times T_{std}}{P_{std} \times T_{dgm}} \right) \qquad (10.13)$$

4. The water vapor volume or mole fraction, B_{H_2O}, is then computed as the volume of water vapor collected, divided by the volume of water vapor plus dry standard gas volume:

$$B_{H_2O} = \frac{V_{wc\,(std)} + V_{wsg\,(std)}}{V_{wc\,(std)} + V_{wsg\,(std)} + V_{dgm\,(std)}} \qquad (10.14)$$

5. Finally, the water vapor volume fraction is used together with the dry molecular weight calculated by Equation 10.8 to find the total molecular weight of the gas, M_{gas}:

$$M_{gas} = M_{dry} \times \left(1 - B_{H_2O}\right) + \left(18 \times B_{H_2O}\right) \qquad (10.15)$$

It is important to note the fact that EPA Method 4 specifies standard temperature as 20 °C (68 °F), not the value usually used for ambient sampling of 25 °C.

Example Problem 10.7. Computing total molecular weight

For the following data, calculate
a) The water vapor fraction, B_{H_2O}
b) The gas total molecular weight, M_{gas}
Dry molecular weight, from Method 3: 29.33
Dry gas meter reading, V_{dgm}: 720 liters
Dry gas meter correction factor, Y: 0.97
Dry gas meter average temperature, T_{dgm}: 26 °C
Dry gas meter average gauge pressure: 45 mm H_2O
Barometric pressure: 746 mm Hg
Volume of water condensed in Method 4: ΔV = 6.5 mL
Mass of water adsorbed on silica gel: Δm = 3.5 g
Density of water, ρ = 0.9982 g/mL

Solution

a) Volume of water vapor condensed at standard temperature (20 °C) and pressure in standard cubic meters (scm) (Equation 10.11):

$$V_{wc(std)} = 8.7 \text{ L}$$

Volume of water vapor adsorbed, scm (Equation 10.12):

$$V_{wsg(std)} = 4.7 \text{ L}$$

Volume of dry gas sampled, standard cubic meters (Equation 10.13):

$$V_{dgm(std)} = 712.3 \text{ L}$$

Using Equation 10.14:

$$B_{H_2O} = (V_{wc} + V_{wsg})/(V_{wc} + V_{wsg} + V_{dgm})$$
(all terms at standard T and P)
$$= 0.0184 \text{ or } 1.84\%$$

b) Equation 10.15 gives:

$$M_{gas} = (1 - 0.0184) \times 29.33 + 0.0184 \times 18$$
$$= 29.12 \text{ g/gmole.}$$

It should not be surprising that the total molecular weight is less than the dry molecular weight in Example Problem 7. Water vapor, with a molecular weight of 18, dilutes the heavier dry gas, resulting in a lighter mixture.

The value for total molecular weight of the gas, as found from Equation 10.15, is to be used in the equation to compute gas velocity for Method 2, Equation 5.7b:

$$v_s = C_p \times K_p \times \left[\frac{(T_{gas} \times \Delta P_p)}{(P_{stat} \times M_{gas})}\right]^{\frac{1}{2}}$$

10.7 PSYCHROMETRY

Some alternative approximation methods to compute the water vapor fraction in a gas mixture are authorized for preliminary emission evaluations in EPA Method 4. These generally involve psychrometry — the use of wet bulb and dry bulb thermometers and a psychrometric chart or equation to determine water vapor partial pressure and volume fraction.

NOMENCLATURE FOR PSYCHROMETRY:

P_{H_2O}: Partial pressure of water vapor in a mixture
P_{sat}: Saturation partial pressure of water vapor (P_{sat} is the maximum value of P_{H_2O}; depends on temperature)
P_{stat}: Absolute gas pressure
T_{dp}: Dew point, temperature at which condensation of water vapor from the air occurs
T_{db}: Dry bulb gas temperature; the measured gas temperature

T_{wb}: Wet bulb temperature; gas temperature which accounts for the evaporation of water into the gas and hence the gas's relative humidity

Φ: Relative humidity; P_{H_2O}/P_{sat}

Three alternatives to Method 4 will be discussed.

1. Saturated gas streams. If the gas contains the maximum possible moisture at that temperature, $T_{wb} = T_{db}$. The saturation partial pressure of water vapor, P_{sat}, can be obtained from a saturated steam table at the gas temperature. Because, for ideal gases, the partial pressure ratio is equal to the volume fraction, and thus,

$$B_{H_2O} = \frac{P_{sat}}{P_{stat}} \tag{10.16}$$

Saturation pressure tables are found in Appendix G of this text. The value of B_{H_2O} may then be used to calculate total molecular weight as described above.

2. Unsaturated gas streams; psychrometric chart. If T_{wb} is less than T_{db}, the gas stream is not saturated ($\Phi < 100\%$), as is usually the case. A psychrometric chart like that in Appendix G can be used to look up Φ and T_{dp}, given T_{db} and T_{wb}. The chart must be appropriate for the gas pressure, P_{stat}, however. As for the saturated case, a saturated steam table can be used to find P_{sat} at T_{db} and P_{H_2O} is determined from

$$P_{H_2O} = \Phi \times \left(P_{sat}/100\right) \tag{10.17}$$

Then,

$$B_{H_2O} = P_{H_2O}/P_{stat} \tag{10.18}$$

High-temperature psychrometric charts can be formulated to give values for B_{H_2O} directly. See Appendix G for an example.

3. Unsaturated gas streams; the Carrier equation. There are two versions of the same basic equation, both of which are equations which approximate the psychrometric chart curves and account for pressure differences. The information needed is the gas stream T_{db} and T_{wb}, the absolute gas pressure P_{stat}, and the saturation vapor pressure at the wet bulb temperature, $P_{sat,w}$. Then, the Carrier equation is given by:

$$P_{H_2O} = P_{sat,w} - \frac{\left(P_{stat} - P_{sat,w}\right) \times \left(T_{db} - T_{wb}\right)}{2800 - \left(1.3 \times T_{wb}\right)} \tag{10.19}$$

or

$$P_{H_2O} = P_{sat,w} - \left(3.67 \times 10^{-4}\right) \times P_{stat} \times \left(T_{db} - T_{wb}\right) \times \left(1 + \frac{\left(T_{wb} - 32\right)}{1517}\right) \tag{10.20}$$

Note that these equations are formulated to accept units of degrees fahrenheit for temperatures and inches of mercury for pressures only. Equation 10.18 can be used to compute B_{H_2O}.

Example Problem 10.8. Finding B_{H_2O} by Psychrometry

Find B_{H_2O} a) by use of psychrometric chart, and b) by both forms of the Carrier equation for the following conditions:

$$T_{db} = 96° \text{ F}$$
$$T_{wb} = 82 \text{ °F}$$
$$P_{stat} = 29.53'' \text{ Hg}$$

Solution

a) The psychrometric chart of Appendix G gives $\Phi = 55\%$. The saturated steam table gives $P_{sat} = 1.712''$ Hg at $T_{db} = 96$ °F. Equation 10.17 gives $P_{H_2O} = 0.942''$ Hg.

b) The saturated steam table of Appendix G may be used to look up $P_{sat,w}$ at $T_w = 82$ °F; $P_{sat,w} = 1.102''$ Hg.

Equation 10.19 gives $P_{H_2O} = 0.954''$ Hg, while Equation 10.20 gives $P_{H_2O} = 0.945$.

Equation 10.18 gives for the three values of B_{H_2O}, respectively:

$$B_{H_2O} = 0.0319 \text{ or } 3.19\%$$
$$B_{H_2O} = 0.0323 \text{ or } 3.23\%$$
$$B_{H_2O} = 0.0320 \text{ or } 3.20\%, \text{ excellent agreement.}$$

Example Problem 10.9. Comprehensive example.[3] Reprinted with permission from *Engineering Thermodynamics*, 2nd ed., Burghardt, M.D., Harper Collins.

A coal-fired utility boiler uses fuel that contains 75% carbon and 8% ash. When burned, the fuel leaves no unburned residue. Air enters the unit at T = 32 °C and 80% relative humidity. The gas at the stack exit has $T_{gas} = 360$ °C. Orsat analysis of the exhaust gas gives: 12.6% CO_2, 6.2% O_2, and 1% CO. Find:

a. Air-to-fuel ratio
b. Percent excess air, % EA
c. Volume of exhaust gas (m^3) per kg of fuel
d. Volume of air (m^3) per kg of fuel
e. Dewpoint
f. Dry molecular weight of the exhaust gas
g. Total molecular weight of the exhaust gas
h. Percent by weight of carbon and hydrogen in the coal

Solution

The balanced equation is:

$$\left[C_xH_y\right] + a * \left[O_2\right] * 3.76 * a * \left[N_2\right] + b * \left[H_2O\right] \rightarrow$$

$$12.6\left[CO_2\right] + 6.2\left[O_2\right] + 1.0\left[CO\right] + 80.2\left[N_2\right] + d * \left[H_2O\right]$$

Solving: $x = 13.6$

$a = 80.2/3.76 = 21.33$

moles dry air = 21.33 + 80.2 = 101.53 per mole fuel

From the psychrometric chart in Appendix G, at T = 32 °C (90 °F), and relative humidity = 80%, find humidity ratio = 0.0244 = m_{H_2O}/m_{air}. Now, $(m_{H_2O}/m_{air}) \times (M_{air}/M_{H_2O}) = 0.0244 \times (28.96/18) = 0.03927$ moles H_2O per mole air.

Since there are 101.53 moles air/mole fuel, 0.03927 × 101.53 = 3.987 moles H_2O per mole fuel. This is "b" in the combustion equation. Next, use oxygen balance: 21.33 × 2 + 3.987 = 12.6 × 2 + 6.2 × 2 + 1 + d to obtain d = 8.04.

Finally, hydrogen balance: $y + 2 \times b = 2 \times d$ $y = 8.11$, and $C_{13.6}H_{8.11}$ is the fuel.

a)

$$\frac{A}{F_{act}} = \left(\frac{21.33 \times 4.76}{1}\right) \times \left(\frac{28.96}{13.6 \times 12 + 8.11}\right) = 17.16\frac{\text{kg air}}{\text{kg fuel}}$$

The stoichiometric equation is:

$$C_{13.6}H_{8.11} + (a \times O_2) + (3.76 \times a \times N_2)$$
$$= (13.6 \times CO_2) + (4.06 \times H_2O) + (3.76 \times a \times N_2)$$

Balancing O_2: a = 15.625, and (3.76 × a) = 58.75

Therefore,

$$\frac{A}{F_{stoich}} = \left(\frac{15.625 \times 4.76}{1}\right) \times \left(\frac{28.96}{13.6 \times 12 + 8.11}\right) = 12.57$$

where the denominator is $171.3 = M_{fuel}$, the molecular weight of the fuel, and the percent excess air is:

b)

$$EA\% = \left[\frac{A/F_{act}}{A/F_{stoich}} - 1\right] \times 100 = 36.5\%$$

c) Moles exhaust gas/mole fuel = 12.6 + 6.2 + 80.2 + 8.04 + 1.0 = 109.04 moles.

For 1 mole fuel, the exhaust gas volume at pressure P_{ex} = 101.3 kPa is:

$$V_{ex} = \frac{108.04 \times R_u \times T_{ex}}{P_{ex}} = 5611.7 \frac{m^3}{\text{kgmole fuel}}$$

and V_{ex} per kilogram fuel = 5611.7/171.3 = 32.76 m³/kg of the combustible portion of the fuel.

d) Similarly, V_{air} per kg of fuel =

$$V_{air} = \left[\frac{(21.33 * 4.76) * R_u * T_{in}}{P_{in}}\right] * \frac{1}{M_{fuel}} = 14.83 \frac{m^3}{\text{kg fuel}}$$

e) $T_{dp} = T_{sat}$ at the partial pressure of water in the exhaust gas, p_{H_2O}.

$$\frac{p_{H_2O}}{P_{TOT}} = B_{H_2O} = \text{mole fraction} = \frac{8.04}{108.04} = 0.074$$

Therefore, p_{H_2O}= 0.074 × 101 = 7.52 kPa or 56.4 mm Hg, and from steam tables (See Appendix G), $T_{dp} = T_{sat}$ = 40 °C

f) $M_{dry} = \Sigma B_i * M_i = (12.6 * 44 + 6.2 * 32 + 1 \times 28 + 80.2 * 28)/100 = 30.26$

g) $M_{TOT} = (1 - B_{H_2O} \times M_{dry} + B_{H_2O} \times 18$
$= (1 - 0.074) \times 30.26 + 0.074 \times 18 = 29.35$

h) M_{fuel} = 171.3, so

$$\%C \text{ by wt} = \frac{n_c \times M_c}{n_{fuel} \times M_{fuel}} = \frac{13.6 \times 12}{1 \times 171.3} \times 100 = 95.3\%$$

$$\%H \text{ by wt} = \frac{n_H \times M_H}{n_{fuel} \times M_{fuel}} = \frac{8.11 \times 1}{1 \times 171.3} \times 100 = 4.7\%$$

CHAPTER 10 PROBLEMS

1. An Orsat analysis of air yields:
   $CO_2 = 0.03\%$
   $O_2 = 20.6\%$
   $CO = 0\%$
   By another technique, it is found that % argon = 1.04.
   Assuming the remainder is nitrogen, compute the dry molecular weight.
2. Orsat results for exhaust from a boiler are:
   $CO_2 = 10.1\%$
   $O_2 = 8.3\%$
   $CO = 0.9\%$
   a) Find the dry molecular weight.
   b) Find the percent excess air.
   c) If the moisture percent is 4.96%, compute the total molecular weight.
   d) Compute the air-to-fuel ratio on a mass basis.
3. Oak logs are burned in a wood stove. The wood can be assumed to be composed of molecules of the form $C_AH_BO_C$, where an ultimate analysis reveals A = 42.17, B = 53.0, C = 13.98. The combustion equation is

$$d[C_AH_BO_c] + e[O_2] + 3.76 \times e[N_2]$$

$$= f[CO_2] + g[CO] + h[O_2] + 3.76 \times e[N_2] + i[H_2O]$$

   Orsat readings were 6.8% CO_2, 12.4% O_2, and 1.4% CO. (Use the concept of balancing the combustion equation for 100 moles of dry products.)
   Find A/F and %EA: note that there is oxygen in the fuel, so you cannot find the %EA directly from the Orsat data.
4. The following data were obtained for an unknown fuel C_xH_y burned in with dry air:
   $\%O_2 = 5.4$
   $\%CO_2 = 12.3$
   $\%CO = 0.7$
   Dry gas meter reading = 55 ft^3
   Dry gas meter temp = 65 °F
   Dry gas meter absolute pressure = 29.6" Hg
   Dry gas meter Y value = 1.035
   a) Balance the combustion equation.
   b) Compute B_{H_2O} in the exhaust gas using the balanced equation.
   c) Compute dry molecular weight of exhaust gas.
   d) Compute total molecular weight of exhaust gas.

e) Compute the volume (m^3) of exhaust gas at standard temperature and pressure per kilogram of fuel.

5. A circular exhaust duct that has a diameter of 45" is to be tested for velocity and flow rate. There is a convenient sampling location which is 22 ft downstream from a bend in the duct. The sampling location is 10 ft upstream from another flow disturbance. It is proposed to do velocity measurement at this location.
   a) How many sampling points must be used, according to EPA Method 1? (Table 10.1).
   b) Sketch the duct cross-section and show where the points are to be located.
   c) Beside the sketch, make a table of all points giving the distance in inches from the inner duct wall to each point.
6. Find B_{H_2O} by use of
   a) Psychrometric chart and
   b) Both forms of the Carrier equation, for the following conditions:
      T_{db} = 175 °F
      T_{wb} = 110 °F
      P_{stat} = 29.92" Hg
7. Stack sampling data:
   DGM reading = 5.3 ft^3
   $C_{dgm} = Y_{avg}$ = 1.08
   $T_{dgm,avg}$ = 12 °C
   ΔH_{avg} for orifice = 1.36" H_2O
   P_{bar} = 29.86" Hg
   Liquid water condensed = 4 mL
   Water adsorbed onto silica gel = 2.6 g
   Orsat: CO_2 = 10.1%
   O_2 = 8.3%
   CO = 0.9%
   Calculate: a) B_{H_2O}
   b) Dry molecular weight
   c) Total molecular weight
8. Given the following psychrometry data:
   Stack gas temperature = 240 °F
   Wet bulb temperature = 140 °F
   P_{bar} = 29.9" Hg
   Stack gas pressure: $P_{s,\,gage}$ = −3.4" H_2O
   Calculate: a) B_{H_2O}
   b) Relative humidity
   c) Dewpoint
9. A rectangular duct 2.4 × 2 ft is to be sampled for both particulates and nonparticulates. In order to meet the criterion for the minimum number of traverse points (12), how far should the measurement site be constructed downstream (B) from the nearest disturbance?

11 EMISSION TEST EQUIPMENT CALIBRATION

11.1 INTRODUCTION

Source testing techniques for quantifying emissions of air pollutants are carefully documented procedures that integrate and require a good understanding of most of the preceding material in this book. While Method 5 is the most fundamental of the EPA Source Testing Methods, many of the same techniques are incorporated into other pollutant emission measurement procedures:[31] e.g., EPA Methods 6, 7, 8, 10–20, and 25, found in 40CFR60,[2] and SW-846 Methods 0010 (Modified Method 5) and 0030 (Volatile Organic Sampling).[1,17] Previously discussed processes involved in any emission determination include:

Basic gas concepts (Chapter 2)
Volume measurement techniques (Chapter 3)
Flow rate measurement techniques (Chapter 4)
Velocity measurement techniques (Chapter 5)
Flow moving and controlling devices (Chapter 6)
EPA Methods 1–4 (Chapter 10)

Raw data to be measured for any pollutant emission quantification include:

Gas flow rate and velocity (Methods 1 and 2)
Molecular weight (Method 3)
Moisture (Method 4)
Temperature and pressure
Gas sample volume

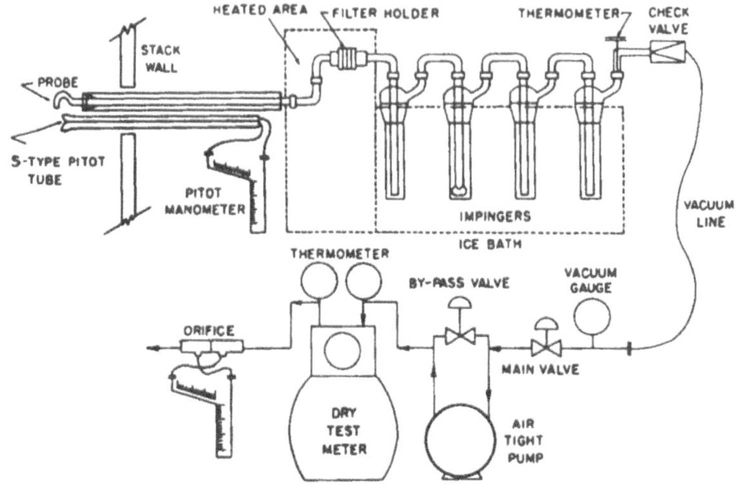

Figure 11.1. Schematic of EPA Method 5 equipment (Reference 12).

Stack or duct cross-section
Mass of pollutant collected

The ultimate objective of any air pollutant emissions test is an accurate and defensible evaluation of the source's particulate emission rate in the same units as the pertinent regulation. From the raw data, the following values are to be calculated:

Volume flow rate through the duct or stack in dry standard cubic meters or feet per hour (dscmh or dscfh)
Concentration of pollutant in the gas (ppm, ppb, g/dscm or lb/dscf)
Mass emission rate in g/sec or lb/hr
Emission rate in some units of mass per unit of process measurement (e.g., pounds of pollutant per million Btu of fuel used)

Chapter 12 will focus on integrating the measurements into a complete emissions test for any air pollutant and calculating the necessary parameters to evaluate the emission rate and document the accuracy of the test. This chapter deals with the complex process of preparing and calibrating equipment for an emissions test.

11.2 CALIBRATION OF EQUIPMENT

Figures 11.1 and 11.2 show the equipment generally used for EPA Method 5 particulate emission assessment. Method 5 is formally entitled "Determination

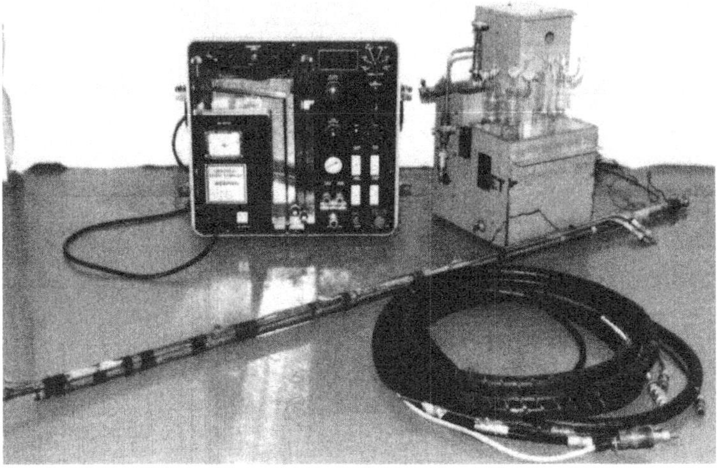

Figure 11.2. EPA Method 5 sampling equipment.

of Particulate Emissions from Stationary Sources". There are a number of manufacturers whose equipment varies slightly but performs all the same essential functions. The requirements for materials and construction details are spelled out in 40CFR60, Appendix A, Method 5[2] and in APTD 0576, "Maintenance, Calibration and Operation of Isokinetic Source Sampling Equipment".[13,25]

Prior to use, an extensive amount of equipment calibration is necessary. The sampling nozzle diameter must be accurately measured; the pitot tube must be calibrated as described in Chapters 5 and 10; the dry gas meter (for sample volume measurement) and the orifice meter (for sampling rate monitoring), which are housed in the meter box, must be calibrated as described in Sections 3.8 and 4.3, respectively; the heater controls, temperature sensors, and pressure measuring devices must all be calibrated; and all flow paths must be shown to be leak-free. Though covered in Chapters 3 and 4, the process for calibration of the dry gas meter and orifice meter bears further discussion.

11.3 DRY GAS METER — Y_{AVG}

As shown in Figure 11.3, the dry gas meter and orifice meter are calibrated against a calibrated wet test meter. The data specified in Figure 11.4 are recorded at a number of different flow rates as indicated by the orifice readings, ΔH, spanning the value expected during operation of the sampling system. A reasonable range for ΔH values is 10 to 100 mm H_2O (0.4 to 4" H_2O). For each flow

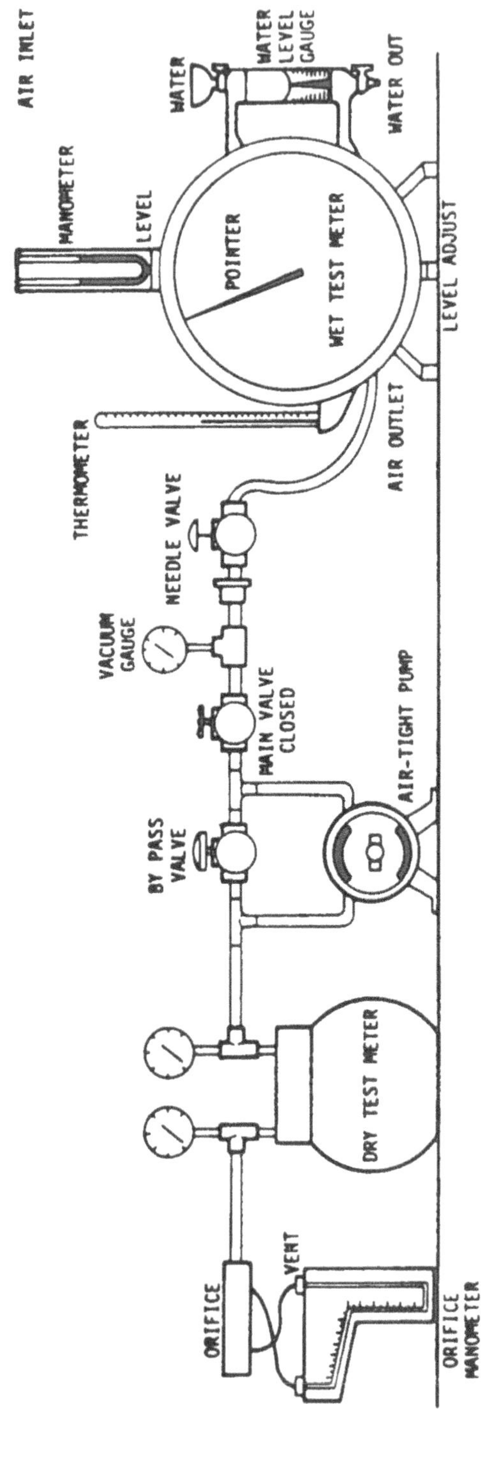

Figure 11.3. Dry gas meter and orifice meter (Reference 12).

METER BOX CALIBRATION DATA AND CALCULATION FORM

(Metric units)

Date ____________ Meter box number ____________

Barometric pressure, P_b = ________ mm Hg Calibrated by ____________

Orifice manometer setting (ΔH), mm H_2O	Gas volume: Wet test meter (V_w), m^3	Gas volume: Dry gas meter (V_d), m^3	Temperatures: Wet test meter (t_w), °C	Temperatures: Dry gas meter: Inlet (t_{d_i}), °C	Temperatures: Dry gas meter: Outlet (t_{d_o}), °C	Temperatures: Dry gas meter: Avg[a] (t_d), °C	Time (θ), min	Y_i	$\Delta H@_i$, mm H_2O
10	0.15								
25	0.15								
40	0.30								
50	0.30								
75	0.30								
100	0.30								
							Avg		

ΔH, mm H_2O	$\frac{\Delta H}{13.6}$	$Y_i = \frac{V_w P_b (t_d + 273)}{V_d (P_d + \frac{\Delta H}{13.6})(t_w + 273)}$	$\Delta H@_i = \frac{0.00117 \Delta H}{P_b (t_d + 273)} \left[\frac{(t_w + 273)\theta}{V_w}\right]^2$
10	0.7		
25	1.8		
40	2.94		
50	3.68		
75	5.51		
100	7 35		

[a] If there is only one thermometer on the dry gas meter, record the temperature under t_d.

METER BOX CALIBRATION DATA AND CALCULATION FORM (metric units)

Nomenclature:

V_w = Gas volume passing through the wet test meter, m^3.

V_d = Gas volume passing through the dry gas meter, m^3.

t_w = Temperature of the gas in the wet test meter, °C.

t_{d_i} = Temperature of the inlet gas of the dry gas meter, °C.

t_{d_o} = Temperature of the outlet gas of the dry gas meter, °C.

t_d = Average temperature of the gas in the dry gas meter, obtained by the average of t_{d_i} and t_{d_o}, °C.

ΔH = Pressure differential across orifice, mm H_2O.

Y_i = Ratio of accuracy of wet test meter to dry gas meter for each run. Tolerance Y_i = Y $\pm$0.02 Y.

Y = Average ratio of accuracy of wet test meter to dry gas meter for all six runs. Tolerance Y = Y $\pm$0.01 Y.

$\Delta H@_i$ = Orifice pressure differential at each flow rate that gives 0.021 m^3 of air at standard conditions for each calibration run, mm H_2O. Tolerance $\Delta H@_i$ = $\Delta H@$ $\pm$3.8 mm H_2O (recommended).

$\Delta H@$ Average orifice pressure differential that gives 0.021 m^3 of air at standard conditions for all six runs, mm H_2O. Tolerance $\Delta H@$ = 46.74 $\pm$6.3 mm H_2O (recommended).

θ - Time of each calibration run, min.

P_b - Barometric pressure, mm Hg.

Figure 11.4. Raw data sheet for DGM/orifice calibration (Reference 12).

rate or ΔH setting, the value of Y or C_{dgm}, the dry gas meter correction factor, is computed from Equation 3.11 and entered in the data sheet (Figure 11.4).

$$C_{dgm} = Y = \frac{V_{wtm} \times C_{wtm} \times P_{bar} \times \left(T_{dgm} + 460\right)}{V_{dgm} \times \left(P_{bar} + \frac{\Delta H}{13.6}\right) \times \left(T_{wtm} + 460\right)} \tag{3.11}$$

The values of Y will vary somewhat with flow rate. If none of the individual values is greater than ±0.02 from the average for all data, and if this average Y is between 0.95 and 1.05, the data are acceptable. The average value of Y will then be employed to correct all dry gas meter readings obtained during sampling to the actual volume.

11.4 ORIFICE METER — ΔH@

The orifice calibration is performed as in Section 4.3, except that instead of developing a curve or plotted line that relates flow rate to the square root of orifice pressure drop, a single coefficient or characteristic that describes the orifice calibration is computed. That parameter is called ΔH@ and is the orifice pressure drop or head loss that corresponds to a flow rate through the orifice of 0.021 scmm (0.75 scfm). This flow rate is optimum for the standard impingers and other apparatus in the Method 5 sampling train. Referring to Equation 4.5,

$$Q_m = K_m \times \left[\frac{T_i}{P_i \times M} \times (\Delta H)\right]^{\frac{1}{2}} \tag{4.5b}$$

this equation can be said to depict a straight-line plot of flow rate through the orifice (Q_m) vs square root of orifice head loss, $(\Delta H)^{1/2}$, for a particular temperature, pressure, and molecular weight for an orifice with coefficient K_m.

ΔH@ and $Q_{m,std}$ (= 0.021 standard cubic meters per minute of air, where T_{std} = 20 °C, P_{std} = 760 mm Hg, M_{air} = 28.96), then, represent the coordinates of one point on a standard calibration plot of orifice head loss vs flow rate for the stack sampler orifice. Squaring Equation 4.5, the relationship between the orifice coefficient and ΔH@ is easily found:

$$Q_m^2 = (0.021)^2 = K_m^2 \times \left(\frac{293}{760 \times 28.96}\right) \times \Delta H@$$

or

$$\Delta H@ = 0.0331 / K_m^2 \quad \text{(SI)} \quad \text{or similarly,}$$

$$\Delta H@ = 0.9244 / K_m^2 \quad \text{(US)} \tag{11.1}$$

These relationships indicate that $\Delta H@$ is as unique a characteristic of the orifice as the orifice coefficient K_m. Since the orifice curve is linear, $\Delta H@$ can be calculated given any other values of ΔH and $Q_{m,std}$ on the calibration line. The meter box orifice calibration process consists of utilizing the same data collected in the dry gas meter calibration (see Figure 11.4) and computing values of $\Delta H@$.

In the following equations, θ is the time for the flow of a volume of air V_{wtm} through the calibration standard.

$$\Delta H@ = 0.00117 \times \frac{\Delta H}{P_{bar} \times T_i} \times \left(\frac{T_{wtm} \times \theta}{V_{wtm} \times C_{wtm}} \right)^2 \text{ (SI)} \qquad \textbf{(11.2a)}$$

$$\Delta H@ = 0.0317 \times \frac{\Delta H}{P_{bar} \times T_i} \times \left(\frac{T_{wtm} \times \theta}{V_{wtm} \times C_{wtm}} \right)^2 \text{ (US)} \qquad \textbf{(11.2b)}$$

The average value of $\Delta H@$ will then be employed to compute flow rates during use of the equipment and to ensure that the proper sampling rate is maintained. For orifices typically used in Method 5 equipment, $\Delta H@$ is normally 46.7 ± 6.4 mm H_2O (1.84 ± 0.25" H_2O).

11.5 DERIVATION OF EQUATION 11.2a

Repeating the orifice equation:

$$Q_m = K_m \times \left[\frac{T_i}{P_i \times M_{air}} \times (\Delta H) \right]^{\frac{1}{2}} \qquad \textbf{(4.5b)}$$

where T_i and P_i are the operating conditions of the gas in the orifice and Q is computed from wet test meter volume, V_{wtm} (see Figure 11.3), and corrected to the conditions of the orifice using Boyle's and Charles' Laws (Equation 2.12),

$$Q_m = \frac{V_{wtm} \times C_{wtm}}{\theta} \left(\frac{P_{wtm}}{P_i} \times \frac{T_i}{T_{wtm}} \right) \qquad \textbf{(11.3)}$$

As shown above, a point on the calibration plot for the orifice is:

$$Q_{std} = 0.021 = K_m \times \left(\frac{T_{std}}{P_{std} \times M_{air}} \times \Delta H@ \right)^{\frac{1}{2}} \qquad \textbf{(11.4)}$$

Equating Equation 11.3 to Equation 4.5b, and dividing the resulting equality into Equation 11.4, then squaring both sides and collecting terms yields:

$$\frac{(0.021)^2\theta^2}{V_{wtm}^2 \times C_{wtm}^2} \times \left(\frac{P_i}{P_{wtm}} \times \frac{T_{wtm}}{T_i}\right)^2 = \frac{\Delta H@}{\Delta H} \times \left(\frac{P_i}{P_{std}} \times \frac{T_{std}}{T_i}\right) \qquad (11.5)$$

With the assumption that both the wet test meter pressure and the orifice operating pressure are nearly equal to barometric pressure, $P_w \approx P_i \approx P_{bar}$:

$$4.41 * 10^{-4}\left(\frac{\theta}{V_{wtm} \times C_{wtm}}\right)^2 \left(\frac{T_{wtm}}{T_i}\right)^2 = \frac{\Delta H@}{\Delta H}\left(\frac{P_{bar}}{P_{std}}\frac{T_{std}}{T_i}\right) \qquad (11.6)$$

and with P_{std} = 760 mm Hg, T_{std} = 293 K:

$$\Delta H@ = 0.00117 \frac{\Delta H}{P_{bar}\ T_i}\left(\frac{T_{wtm}\theta}{V_{wtm}\ C_{wtm}}\right)^2 \qquad (11.2a)$$

12 EPA METHOD 5 PROCEDURES

12.1 INTRODUCTION

EPA Method 5 is entitled "Determination of Particulate Emissions from Stationary Sources". In recent years, a number of variations of Method 5 for particular industries have been promulgated. The 1991 edition of 40CFR60[2] lists, in addition to the basic method, Methods 5A through 5H. Obviously, some very specific and esoteric requirements are spelled out.

As noted in the preceding chapter, the ultimate objective of an emissions test for any air pollutant is an accurate and defensible quantification of the source's emission rate for that pollutant. Because sampling for particulate matter is in many ways more complicated than for a gaseous pollutant, EPA Method 5 has a number of unique requirements designed to ensure that a representative sample is extracted from the duct or stack. One of the most significant of these is the requirement for isokinetic sampling.

12.2 THE NEED FOR ISOKINETIC SAMPLING

The sample of exhaust gas extracted from the duct or stack must be a "representative" sample; that is, it must be a small portion of the gas exhausted that has all the same average characteristics as the gas that is routinely exhausted. The velocity, temperature, and pollutant concentration of the gas may vary with time and with location in the cross-sectional area of the duct. To deal with spatial variations, Method 1 requires that portions of the sample be

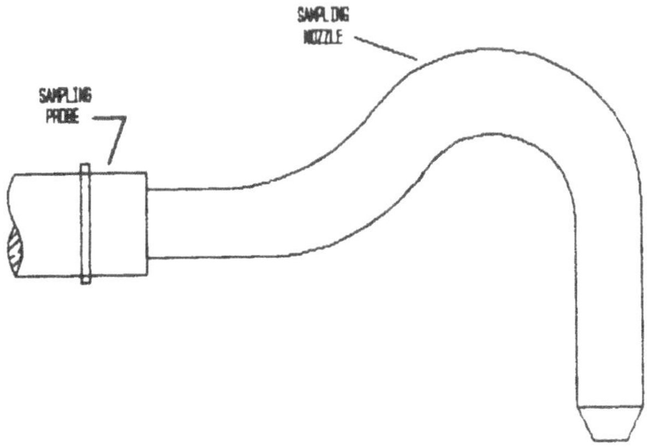

Figure 12.1. Nozzle.

extracted from a number of different locations in the cross-section. At each of these locations, the velocity of the gas being drawn into the end of the sampling nozzle, v_n, must be matched to that of the gas passing by the nozzle in the stack or duct, v_s. (See Figure 12.1) The isokinetic requirement is:

$$v_n = v_s \tag{12.1}$$

If the velocities are not equal, the gas flow lines will be disturbed around the tip of the nozzle. If the particulate matter suspended in the gas stream consists of a range of sizes of particles, the smaller, lighter ones will tend to follow the gas flow lines, while the larger ones, due to their greater inertia, will not bend with the gas flow lines, but continue on straighter paths. The consequence of anisokinetic sampling is depicted in Figure 12.2.

- $v_n = v_s$: all size particles follow flow lines and sample is representative of gas in stack.
- $v_n > v_s$: more gas is drawn into nozzle, but larger particles do not follow the gas, so less particulate mass is collected per volume of gas, and the concentration measured is nonrepresentative: lower than in the stack.
- $v_n < v_s$: less gas is drawn into the nozzle, but larger particles do not follow the gas around the nozzle — they enter it. More particulate mass is collected per volume of gas, and the concentration measured is nonrepresentative: higher than in the stack.

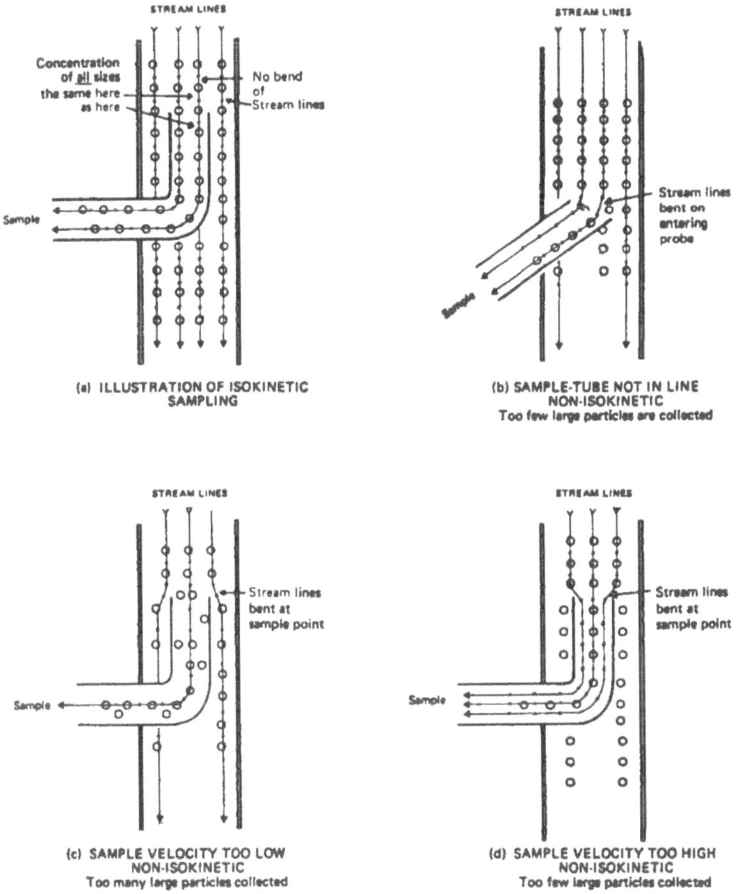

Figure 12.2. Anisokinetic sampling. Reprinted with permission from *Air Sampling and Analysis*, Lodge, J. P., CRC Press, Boca Raton, FL.

It is clear that the sampling nozzle velocity must remain equal to the duct or stack gas velocity at all times, in order to ensure that the sample of gas removed is representative. The gas velocity, however, may change when the probe is moved, or may vary with time; thus, real-time nozzle and stack gas velocity measurements must be available to allow the emission tester to make adjustments in the sample flow rate.

12.3 VELOCITY INDICATORS

Stack gas velocity is measured with a pitot tube and is proportional to $(\Delta P_p)^{1/2}$ as shown in Chapter 5 by Equation 5.7b.

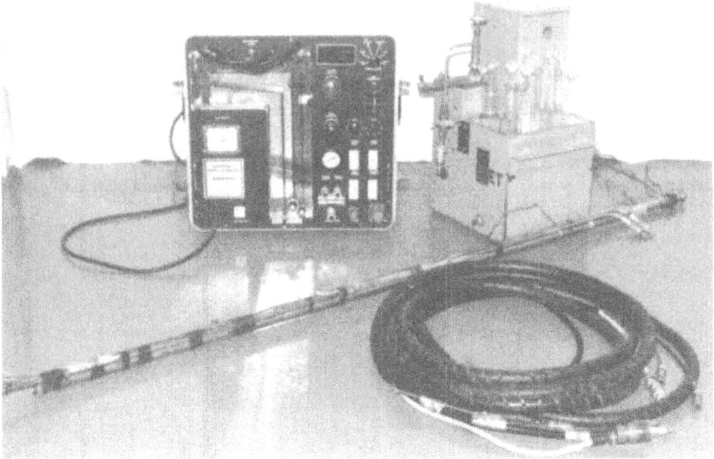

Figure 12.3. Emission sampler control box. (Courtesy Andersen Samplers.)

$$v_s = C_p \times K_p \times \left\{\left(T_{gas} \times \Delta P_p\right)/\left(P_{stat} \times M_{gas}\right)\right\}^{\frac{1}{2}} \quad \textbf{(5.7b)}$$

One of a pair of manometers or pressure gauges mounted on the emission sampler control box is connected, through the umbilical cord, to the calibrated s-type pitot and displays ΔP_p. See Figure 12.3.

In Chapter 4, the equation for flow rate through an orifice was developed (Equation 12.2)

$$Q_m = K_m \times \left(\frac{T_m}{P_m \times M_m} \times \Delta H\right)^{\frac{1}{2}} \quad \textbf{(12.2)}$$

Here, the subscript m refers to conditions of the gas in the orifice meter. In Chapter 11, it was shown that the unique orifice characteristic ΔH@ could be substituted for K_m in Equation 12.2 with a constant that depends on the system of units selected:

$$\Delta H@ = 0.0331/K_m^2 \quad \text{(SI)} \quad \text{or similarly,}$$

$$\Delta H@ = 0.9244/K_m^2 \quad \text{(US)} \quad \textbf{(11.1)}$$

Thus, Equation 12.2 becomes:

$$Q_m = \left(\frac{0.0331}{\Delta H@}\right)^{\frac{1}{2}} \times \left(\frac{T_m}{P_m \times M_m} \times \Delta H\right)^{\frac{1}{2}} \quad \text{(SI) or}$$

$$Q_m = \left(\frac{0.9244}{\Delta H@}\right)^{\frac{1}{2}} \times \left(\frac{T_m}{P_m \times M_m} \times \Delta H\right)^{\frac{1}{2}} \quad \text{(US)} \qquad (12.3)$$

Since all of the gas that enters the nozzle is either water vapor removed in the impingers or gas that will pass through the rest of the sampling train and eventually through the orifice (see Figure 11.1), the nozzle velocity, v_n, is proportional to the square root of orifice head, $(\Delta H)^{1/2}$. The other pressure gauge on the meter box shown in Figure 12.3 displays ΔH.

It is the emission tester's job to maintain isokinetic sampling throughout the process. Since the information available to the operator of the meter box is the pitot readout, ΔP_p, and the orifice readout, ΔH, an easily used relationship between these two parameters is needed.

12.4 ISOKINETIC RELATIONSHIP BETWEEN $\Delta H@$ AND ΔP_p

The flow rate through the nozzle, Q_n, is found from the calculated orifice flow rate of Equation 12.3 by making two adjustments:

1. A pressure and temperature correction
2. The addition of the volume of water vapor that flows through the nozzle, but is removed before the gas reaches the orifice

$$Q_n = Q_m \times \frac{P_m}{P_{stat}} \times \frac{T_{gas}}{T_m} \times \left(\frac{1}{1 - B_{H_2O}}\right) \qquad (12.4)$$

The water vapor volume fraction, B_{H_2O}, is measured by Method 4 (see Section 10.6). The velocity through the nozzle is given by Equation 12.5.

$$v_n = \frac{Q_n}{A_n} = \frac{Q_n}{\frac{\pi}{4} D_n^2} \qquad (12.5)$$

where A_n is the nozzle area and D_n is nozzle diameter in units consistent with those of Q_n.

Combining Equations 12.3, 12.4, and 12.5, one obtains Eauation 12.6.

$$V_n = \frac{\left(\frac{0.0331}{\Delta H@}\right)^{\frac{1}{2}} \times \left(\frac{T_m}{P_m \times M_m} \times \Delta H\right)^{\frac{1}{2}} \times \frac{P_m}{P_{stat}} \times \frac{T_{gas}}{T_m} \times \left(\frac{1}{1 - B_{H_2O}}\right)}{\frac{\pi}{4} D_n^2} \quad \text{(SI)}$$

(12.6)

This nozzle velocity is to remain equal to the stack gas velocity, measured by the pitot tube and given as Equation 5.7b. Thus, Equation 12.6 is set equal to Equation 5.7b. The resulting complicated equation can be simplified by squaring both sides and combining terms to yield, in SI units, Equation 12.7.

$$K_p^2 C_p^2 \left(\frac{T_{gas}\, \Delta P_p}{P_{stat}\, M_{gas}}\right) = \frac{\frac{0.0331}{\Delta H@} \left(\frac{T_m}{P_m\, M_m} \Delta H\right) \left(\frac{P_m}{P_{stat}} \frac{T_{gas}}{T_m}\right)^2 \left(\frac{1}{1 - B_{H_2O}}\right)^2}{\left(\frac{\pi}{4}\right)^2 D_n^4}$$

(12.7)

The gas in the orifice is the same as stack gas with water removed, so $M_m = M_d$, the dry molecular weight. Solving for ΔH (= ΔH_{iso}, the orifice head loss that provides isokinetic sampling rate) yields:

$$\Delta H_{iso} = \left[K_p^2 \times \left(\frac{\pi}{4}\right)^2 \times C_p^2\right] \times \left[D_n^4 \times \left(\frac{\Delta H@}{0.0331}\right) \times \left(1 - B_{H_2O}\right)^2 \times \frac{M_d}{M_{gas}} \times \frac{P_{stat}}{P_m} \times \frac{T_m}{T_{gas}}\right] \times \Delta P_p \quad \text{(SI)}$$

(12.8)

For SI,

$K_p = 34.97 \times 3600 = 125{,}892$

$\Delta H@$ in mm H_2O; $\Delta H@/0.0331 = 1/K_m^2$ = unitless

ΔP_p in mm H_2O

P_{stat}, P_m in mm Hg

T_s, T_m in K

M_d, M_{gas} in g/gmole

D_n in meters; or if in mm, a constant of 10^{-12} must be included in Equation 12.8.

For US,

$K_p = 85.49 \times 3600 = 307{,}764$

$\Delta H@$ in inches H_2O; [$\Delta H@/0.9244$, unitless (0.9244 replaces 0.0331 in Equation 12.8)]

ΔP_p in inches H_2O

P_{stat}, P_m in inches Hg

T_s, T_m in °R

M_d, M_{gas} in lb/lbmole

D_n in feet, or if in inches, a constant of 4.82×10^{-5} must be included in Equation 12.8.

Thus, for SI, with nozzle diameter in millimeters,

$$\Delta H_{iso} = \left[8.204 \times 10^{-5} \times D_n^4 \times \Delta H@ \times C_p^2 \times \left(1 - B_{H_2O}\right)^2 \times \frac{M_d}{M_{gas}} \times \frac{T_m}{T_{gas}} \times \frac{P_{stat}}{P_m}\right] \times \Delta P_p \quad \text{(SI)} \qquad \textbf{(12.9)}$$

For Customary US units, with nozzle diameter in inches,

$$\Delta H_{iso} = \left[846.7 D_n^4 \times \Delta H@ \times C_p^2 \times \left(1 - B_{H_2O}\right)^2 \times \frac{M_d}{M_{gas}} \times \frac{T_m}{T_{gas}} \times \frac{P_{stat}}{P_m}\right] \times \Delta P_p \quad \text{(US)} \qquad \textbf{(12.9a)}$$

(A multiplier of 3600 sec/hr is included in both constants to accommodate stack gas velocity, v_s; units of length per second and sampling rate, Q, units of volume per hour.)

In Equations 12.9 and 12.9a, the values for $\Delta H@$ and C_p are found by calibration before the equipment is used. The values for B_{H_2O}, M_d, M_{gas}, P_{stat}, P_m, and T_m can be estimated from preliminary measurements. Therefore, values for these parameters can be incorporated into an overall constant K_n:

$$\Delta H_{iso} = K_n \times \frac{D_n^4}{T_{gas}} \times \Delta P_p \qquad \textbf{(12.10)}$$

Equations 12.9, 12.9a, and Equation 12.10 can be programmed into a portable programmable calculator or spreadsheet software on a portable computer for easy use during emission testing. Should any of the parameters in Equation 12.9 change during testing (in particular, T_{gas} and ΔP_p are quite likely to vary), a new value for ΔH_{iso} can be quickly computed from Equation 12.10 and the sampling rate adjusted to maintain isokinetic sampling. Before the days of programmable calculators and portable computers, nomographs or specially designed slide rules were used for the same purpose.[18]

Example Problem 12.1. Use of spreadsheet for isokinetic sampling rate

Set up a spreadsheet with labeled input cells and Equation 12.9 (US units) to compute ΔH_{iso} for the following data:

$\Delta H@ = 1.7"\ H_2O$
$C_p = 0.80$
$T_m = 50\ °F$
$P_{stat} = 29.7"\ Hg$
$P_m = 27.73"\ Hg$
$B_{H_2O} = 0.22$
$M_d = 29.6$
$M_{gas} = 27.05$
$D_n = 0.5"$
$T_{gas} = 600, 650\ °F$
$DP_p = 0.06, 0.10, 0.13, 0.16"\ H_2O$

(Compute 8 values for ΔH_{iso})

Solution

Table 12.1 Spreadsheet to compute isokinetic sampling rate.

Input Data

Common values

$\Delta H@$	1.7
C_p	0.8
T_m	50.0
P_{stat}	29.7
P_m	27.73
B_{H_2O}	0.22
D_n	0.5
M_d	29.6
M_{gas}	27.05

	Set 1	Set 2	Set 3	Set 4	Set 5	Set 6	Set 7	Set 8
T_{gas}	600	600	600	600	650	650	650	650
ΔP_p	0.06	0.1	0.13	0.16	0.06	0.1	0.13	0.16

Constant $K_n = 846.7 \times \Delta H@ \times C_p^2 \times (1 - B_{H_2O})^2 \times M_d/M_{gas} \times T_m \times P_{stat}/P_m$
$= 335{,}027.9$

$\Delta H_{iso} = K_n \times D_n^4/T_{gas} \times \Delta P_p$

	Set 1	Set 2	Set 3	Set 4	Set 5	Set 6	Set 7	Set 8
ΔH_{iso}	1.185	1.975	2.568	3.161	1.132	1.886	2.452	3.018

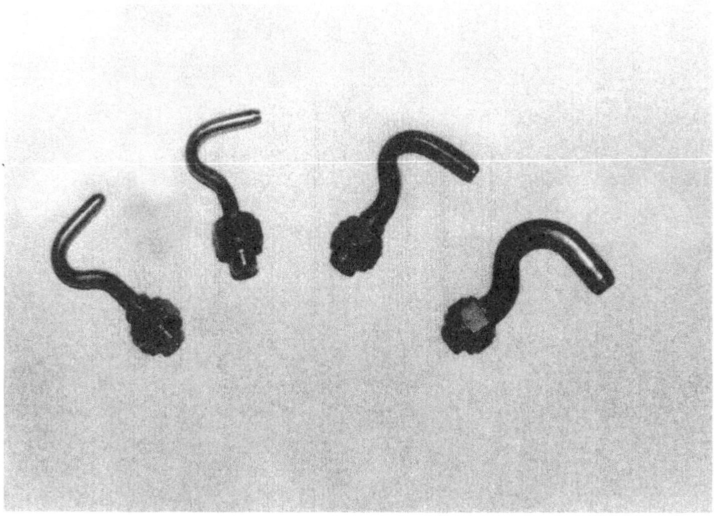

Figure 12.4. Sampling nozzles.

12.5 NOZZLE DIAMETER SELECTION

In Example Problem 1, the nozzle diameter was given as 0.5". Since the nozzle area has a great effect on the sampling velocity, the selection of nozzle diameter is a significant decision in the process of isokinetic sampling. Typically, emission sampling equipment includes a set of four or five interchangeable sharp-edged nozzles with diameters of approximately $^1/_4$, $^5/_{16}$, $^3/_8$, $^1/_2$, and $^3/_4$" (see Figure 12.4). From this set, the best nozzle is the one that allows isokinetic velocity (i.e., sampling nozzle gas velocity equal to the velocity of the gas in the duct) at a sampling gas flow rate approximately equal to the optimal design flow rate of 0.021 scmm (0.75 scfm). Recall from Chapter 11 that ΔH@ was defined as the orifice head loss that corresponds to this flow rate. Obviously, with a limited number of diameters to choose from, it will not be possible to match the sampling rate to the stack gas flow rate at *exactly* the design flow rate, but it is desirable to have ΔH_{iso} be as close to ΔH@ as possible. This may be difficult if the stack gas velocity varies greatly — note the data in Example Problem 1.

The process of nozzle diameter selection involves a different use of Equation 12.9. Known or estimated values for all emission gas parameters and equipment coefficients are substituted into the equation, except for the desired value for D_n. Since the goal is to have sampling flow rate at the optimum level, ΔHiso is set equal to ΔH@. From preliminary stack measurements, the maximum and minimum values for ΔP_p are used in two solutions of the Equation 12.9 to find the range of nozzle diameters. A nozzle with diameter near the middle of this range would be the best choice. Example Problem 2 clarifies the process.

Example Problem 12.2. Nozzle diameter selection

For the data given, calculate the nozzle diameter that allows isokinetic sampling at flow rates that best match the equipment optimum rate of 0.021 scmm.

$\Delta H@ = 2.03"\ H_2O$
$C_p = 0.793$
$T_m = 70°F$
$P_{stat} = -4"\ H_2O$, gage
$P_m = 2"\ H_2O$, gage
with $P_{atm} = 29.1"$ Hg:
$P_s = 28.8"$ Hg
$P_m = 29.24$ "Hg
$B_{H_2O} = 0.06$ or 6%
$M_d = 28.6$
$T_{gas} = 220$ °F
$v_s = 7.5$ ft/sec
$DP_p = 0.08$ to $0.20"\ H_2O$

Solution

Plugging all the given data into Equation 12.9 (US units), and using $\Delta H@ = 2.03$ for ΔH_{iso}, solve for two different values of D_n — one for each of the given ΔP_p values:

For $\Delta P_p = 0.08"\ H_2O$, $D_n = 0.428"$
For $\Delta P_p = 0.20"\ H_2O$, $D_n = 0.340"$

From a standard set of nozzles, it appears that a $^3/_8$" (0.375") nozzle diameter would best approximate the desired flow rate at both ends of the ΔP_p range. Check by plugging all data into Equation 12.9 (US units) again, but this time use $D_n = 0.375$ and solve for ΔH_{iso} for $\Delta P_p = 0.08$, 0.13, and 0.20" H_2O:

For $\Delta P_p = 0.08"\ H_2O$, $DH_{iso} = 1.19$
For $\Delta P_p = 0.13"\ H_2O$, $DH_{iso} = 1.93$
For $\Delta P_p = 0.20"\ H_2O$, $DH_{iso} = 2.97$

These values are all acceptably close to $\Delta H@$.

12.6 VERIFICATION OF ISOKINETIC SAMPLING

Sections 12.4 and 12.5 discussed how to set the sampling rate to isokinetic. As the gas velocity changes with location in the duct or time, there will necessarily be some response time on the part of the equipment operator to reset the sampling velocity to the new isokinetic condition. The objective is to maintain a nearly isokinetic rate at all times, but the specific requirement of Method 5 is an average sampling velocity of within ± 10% of the stack gas velocity. The ratio of nozzle velocity to stack gas velocity (multiplied by 100) is known as the "percent isokinetic".

$$\%I = \frac{v_n}{v_s} \times 100$$
$$90\% < \%I < 110\% \qquad (12.11)$$

To validate the representativeness of the sample of gas removed from the duct, the percent isokinetic must be independently shown to average between 90 and 110%. This is accomplished by computing the average sampling velocity from the volume sampled, the total sampling time, and the nozzle area.

$$V_n = V_m \times \left(\frac{P_m}{P_{stat}} \times \frac{T_{gas}}{T_m}\right) \times \left(\frac{1}{1 - B_{H_2O}}\right) \qquad (12.12)$$

$$Q_n = \frac{V_n}{\theta} \qquad (12.13)$$

$$v_n = \frac{Q_n}{A_n} \qquad (12.14)$$

where V_n is the volume of gas passing through the nozzle
V_m is the volume of gas passing through the dry gas meter
Q_n is the average volume flow rate through the nozzle
θ is the total sampling time
A_n is nozzle area
v_n is the velocity of the gas through the nozzle

Therefore, the nozzle average velocity is:

$$v_n = \frac{V_m \times Y \times \left(\frac{P_m \times T_{gas}}{P_{stat} \times T_m}\right) \times \left(\frac{1}{1 - B_{H_2O}}\right)}{A_n \times \theta} \tag{12.14a}$$

Recalling that the equation for average gas velocity in the stack is:

$$v_s = K_p \times C_p \times \left(\frac{T_{gas}}{P_{stat} \times M_{gas}}\right)^{\frac{1}{2}} \times \Delta P_p \times \left(\Delta P_p\right)^{\frac{1}{2}}_{avg} \tag{5.7b}$$

Then, substituting Equations 12.14 and 5.7b into Equation 12.11 yields:

$$\%I = \frac{v_n}{v_s} \times 100 = \frac{\dfrac{100 \times V_m \times Y \times \left(\frac{P_m \times T_{gas}}{P_{stat} \times T_m}\right) \times \left(\frac{1}{1 - B_{H_2O}}\right)}{A_n \times \theta}}{K_p \times C_p \times \left(\frac{T_{gas}}{P_{stat} \times M_{gas}}\right)^{\frac{1}{2}} \times \left(\Delta P_p\right)^{\frac{1}{2}}_{avg}} \tag{12.15}$$

which simplifies to Equation 12.16.

$$\%I = \frac{100 \times V_m \times YP_m}{\left(1 - B_{H_2O}\right) \times T_m \times A_n \times \theta \times K_p \times C_p} \times \left(\frac{T_{gas} \times M_{gas}}{P_{stat}}\right)^{\frac{1}{2}} \times \left(\Delta P_p\right)^{\frac{1}{2}}_{avg} \tag{12.16}$$

Other forms of the equation may be used if some parameters have been previously calculated (e.g., Equation 12.17),

$$\%I = \frac{100 \times V_m \times YP_m \times T_{gas}}{v_s \times A_n \times \theta \times P_{stat} \times T_m \times \left(1 - B_{H_2O}\right) \times 60} \tag{12.17}$$

where the factor of 60 is included because v_s is generally in feet or meters per second and θ in minutes. Another alternative, from "intermediate data" is given in Equation 12.18.

$$\%I = \frac{100 \times T_{gas} \times V_{m(std)} \times P_{std}}{60 \times T_{std} \times v_s \times \theta \times A_n \times P_{stat} \times \left(1 - B_{H_2O}\right)} \qquad (12.18)$$

The value for %I is to be calculated immediately after the completion of emission testing. If it is outside the allowable range of 90 to 110%, the results of the test are unrepresentative of the actual particulate emissions and are unacceptable.

CHAPTER 12 PROBLEMS

1. Stack sampling nozzle size selection
   Given the following emission source pretest data:

   $\Delta H@ = 1.95"\ H_2O$
   $Y = 0.985$
   $C_p = 0.823$
   $v_s = 6.5$ ft/sec
   $T_{gas} = 320\ °F$
   $M_{dry} = 28.6$
   $B_{H_2O} = 6.0\%$
   $P_{s,gage} = +4.0"\ H_2O$
   $P_{atm} = 29.72"$ Hg
   $P_{dgm} = P_m = 29.87"$ Hg (DGM press)
   $T_{dgm} = T_m = 65\ °F$ (DGM temp)

   Range of ΔP: minimum = 0.05; maximum = 0.25" H_20
   a) Compute the appropriate nozzle diameters, D_n for the minimum and maximum ΔP values.
   b) Select an acceptable and convenient D_n from this range (your choices for nozzle diameters are: 0.125, 0.1875 0.25", 0.375, 0.5, 0.75, ...).
   c) Calculate the proper isokinetic ΔH for the following ΔP values:

$\Delta P_1 = 0.25"\ H_2O$	$\Delta H_1 =$
$\Delta P_2 = 0.16"\ H_2O$	$\Delta H_2 =$
$\Delta P_3 = 0.05"\ H_2O$	$\Delta H_3 =$

2. Below are data from a Method 5 test.
   Use them to calculate:
   a) Total molecular weight
   b) B_{H_2O}
3. Using the data from Problem 2, calculate percent isokinetic.

Stack Test Data for Problems 2 and 3

V_{dgm}	37.2	Dry gas volume, actual ft^3
P_{MASS}	100.3	mg
$(\Delta P_p)^{1/2}_{avg}$	0.14	average of square root of ΔP, $("H_2O)^{1/2}$
P_{atm}	29.29	" Hg
P_{stat}	–.240	Gage press, " H_2O
DIA	0.5	Stack diameter, ft
C_{dgm}	0.935	Y_{avg}, the dry gas meter coefficient
C_P	0.85	Pitot coefficient
T_{dgm}	538	Temperature of gas in DGM, °R
V_C	5.4	Water condensed, mL
m_{sg}	17.3	Mass adsorbed onto silica gel, g
ΔH_{avg}	0.97	Average orifice head loss
CO_2	7.0	%
O_2	15.9	%
CO	0.0	%
T_S	717	°R
D_n	0.5	Nozzle diameter, inch (Area, $A_n = \pi D_n^2/4 = 0.1963/144 = (0.00136\ ft^2)$
FR	5500	Fuel consumption rate, lbm/hr
θ	60	Total test time, min
HHV	150000	Energy content of fuel, Btu/lbm
T_{std}	528	°R
P_{std}	29.92	" Hg

13 COMPUTING EMISSION RATE

13.1 INTRODUCTION

After the percent isokinetic calculation has demonstrated that the average Method 5 sampling velocity is within 10% of the average stack gas velocity (and the emission test is therefore valid), the particulate concentration and mass emission rate are calculated. For gaseous pollutants, there is no need for isokinetic sampling, but the calculations for concentration and emission rate are identical.

13.2 POLLUTANT COLLECTION

The pollutant extracted from the sampled gas is collected in some appropriate medium. For Method 5, the particles are collected on a fiberglass filter located just prior to the point where the gas enters the four impingers. All of the parts of the sampling train before the filter, and the filter itself, are kept heated above approximately 120 °C (250 °F). The heated apparatus keeps the sampled gas above the dewpoint of water and prevents any semivolatile vapors from condensing to liquid or solid before passing through the filter. Only material that was solid or liquid in the stack gas is filtered out in Method 5. A unique complication of Method 5 and its various adaptations is that some of the particles or droplets may impinge on the inner surfaces of the nozzle or sampling probe. This material must be washed out (usually with acetone) and the mass added to that collected on the filter.

In a modification of Method 5 designed to capture semi-volatile organic compounds, an adsorption chamber is added after the filter (Method 0010 in SW 846;[1] Figure 13.1). For volatile organic sampling, where the isokinetic rate is not required, the sampled gas passes directly through an adsorbing Tenax medium, which removes volatiles (Method 0030 in SW 846;[1] see Figure 13.2). Methods 6 and 7 for sulfur dioxide and nitrogen oxides, respectively, arrange to absorb the gases in appropriate liquid reagents. Other methods (e.g., Methods 10A, 12, 26; see 40CFR60, Appendix A[2]) are designed to capture only the pollutant of concern in an appropriate medium from which the total mass collected can be quantified.

For any collected pollutant, the total mass captured, divided by the total volume of gas measured by the sampling train, gives the concentration of pollutant in the stack gas, C_s, in appropriate mass per volume units. Many continuous emission monitors (CEMs) directly measure the concentration of the pollutant of interest in the extracted gas.

13.3 CONCENTRATION CALCULATION

The volume of gas measured by the dry gas meter is the gas volume less water vapor. This volume is corrected to standard temperature and pressure using an adaptation of Equation 2.12; that is, Equation 13.1.

$$V_{m(std)} = V_m \times Y \times \frac{P_m}{P_{std}} \times \frac{T_{std}}{T_m} = V_m \times Y \times \frac{\left(P_{bar} + \frac{\Delta H_{avg}}{13.6}\right)}{T_{m(avg)}} K_1 \tag{13.1}$$

where ΔH_{avg} is the average orifice head loss throughout the test $T_{m(avg)}$ is the test-long average of the dry gas meter temperature, (and each individual T_m value is already the average of the meter inlet and outlet temperatures)

K_1 is 0.3858 for $T_{std} = 293$ K
$P_{std} = 760$ mm Hg
P_{bar} in mm Hg
V_m in m^3
T_m in K
ΔH_{avg} in mm H_2O

or

K_1 is 17.64 for $T_{std} = 528$ °R
$P_{std} = 29.92$" Hg
P_{bar} in " Hg
V_m in ft^3
T_m in °R
ΔH_{avg} in " H_2O

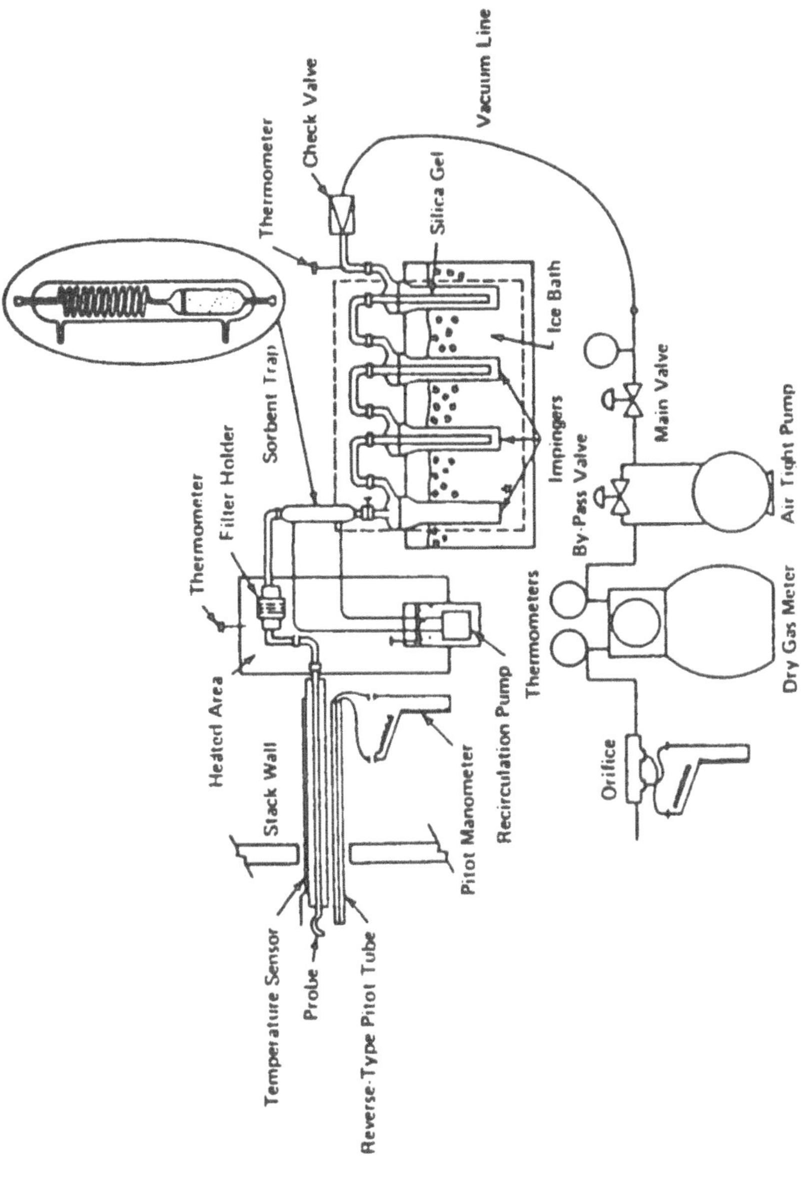

Figure 13.1. Modified Method 5 train (Reference 1).

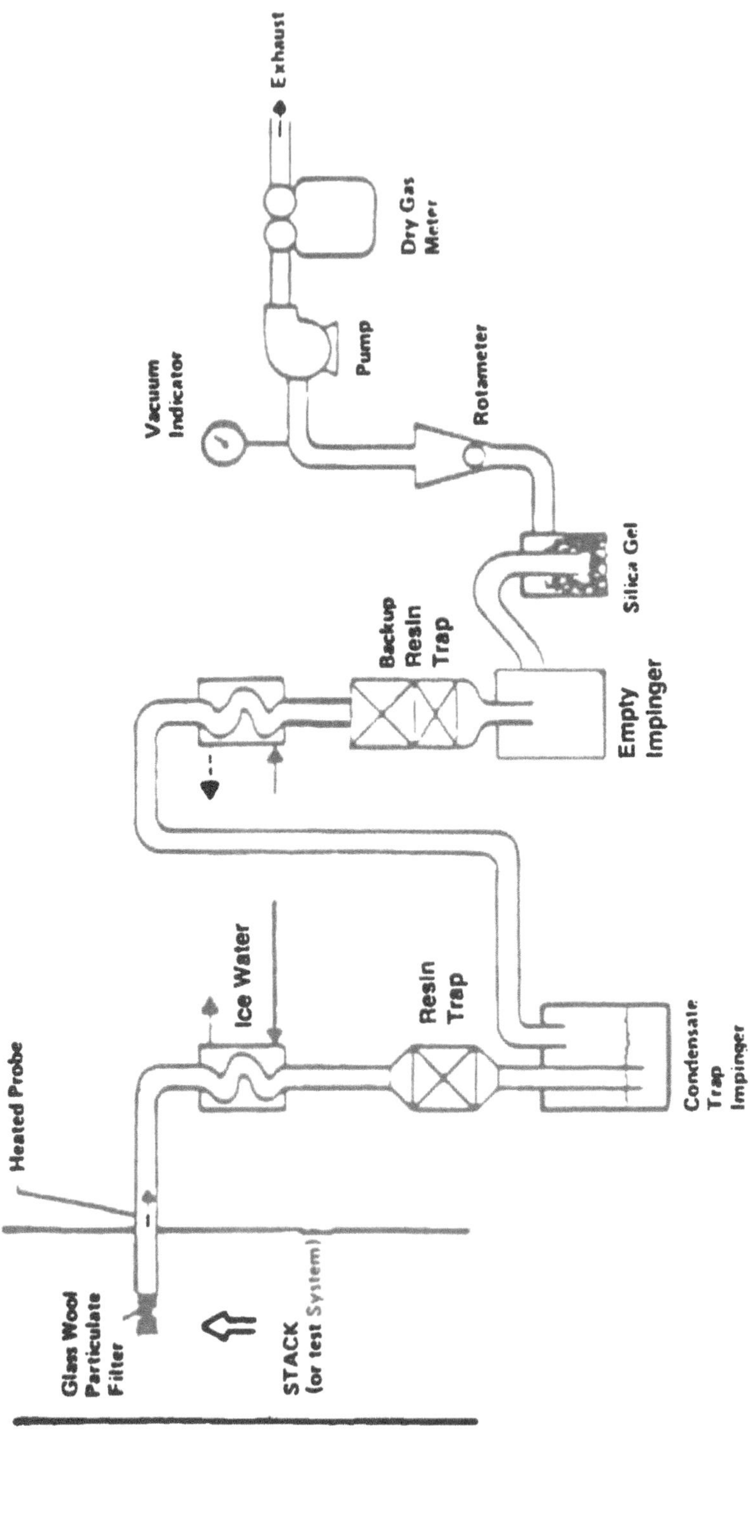

Figure 13.2. Volatile organic sampling train (Reference 1).

The units of $V_{m(std)}$ are dry standard cubic meters (dscm) or dry standard cubic feet (dscf).

The measured concentration of pollutant is:

$$C_s = \frac{m_s}{V_{m(std)}} \qquad (13.2)$$

where m_s is the mass collected and the units for concentration are g/dscm or lb/dscf.

Instrumental analysis methods for some gaseous pollutants (e.g., Method 10 for CO, Method 6C for SO_2, and Method 25 for organic vapors) allow the direct measurement of concentration in the sampled gas in units of ppmv or ppbv. The concentration in mass per standard volume can be calculated using Equation 2.18:

$$\frac{\mu g}{m^3} = 40.9 \times ppmv \times M_{pol} \text{ or}$$
$$\frac{ng}{l} = 40.9 \times 10^{-3} \times ppbv \times M_{pol} \qquad (2.18)$$

While a few air pollutant emission regulations specify a maximum concentration of pollutant in the emitted gas, most set a limit on the pollutant emitted per time or per unit of some production or process parameter. Some examples are listed in Table 13.1.

Table 13.1 Examples of Emission Regulations

1.	Steam power plants	0.03 lb particulate/million Btu
2.	Solid waste incinerators	0.18 g particulate/dscm, corrected to 12% CO_2
3.	Sewage sludge incinerators	0.65 g particulate/kg sludge input
4.	Mercury ore processors	2300 g mercury/24 hr
5.	Nitric acid plants	3 lb NO_2 per ton of acid produced
6.	Portland cement kilns	0.3 lb particulate per ton of solids fed into kiln
7.	Large liquid fossil fuel burning equipment	1.4 g sulfur dioxide/million calories heat input
8.	Plastic parts coating for business machines	1.5 kg VOCs/litercolor coating solids used
9.	New residential wood heaters with catalytic units	4.1 g particulate/hr

13.4 CONCENTRATION CORRECTIONS

Concentration standards, though perhaps the easiest to compare test results to, are not favored because they allow dilution with extra air as a means of reducing the concentration. While actual concentration in the emitted gas might seem the appropriate basis for quantifying the actual rate of pollutant emission, concentration-based regulations are designed for comparisons under standard conditions, so that any pollutant-emitting source can readily be compared to the regulations and to other sources. Different stack temperatures, gas pressures, and different dilutions with excess air could mean actual concentrations bear little relationship to optimum rate of pollutant emission for that type of source.

At a minimum, concentration-based regulations specify maximum allowable pollutant mass per volume of exhaust gas at standard temperature and pressure. More typically, concentration-based regulations overcome the dilution possibility by specifying a correction of reported pollutant concentration to that in gas with a specific CO_2 or O_2 concentration or percent excess air. The solid waste incinerator entry (2) in Table 13.1 is an example of such a standard or regulation.

The pollutant concentration in the exhaust gas calculated from emission test data is:

$$C_s = \frac{m_s}{V_{m(std)}} \tag{13.2}$$

where $V_{m(std)}$ is the measured dry gas volume, corrected to standard temperature and pressure, and C_s is therefore the mass per *dry* standard volume.

Actual concentration

Though usually not needed, the actual concentration can be found from this value by adjusting the volume (and therefore the concentration) for temperature, pressure, and water vapor content of the exhaust gas.

$$C_{s(act)} = C_s \times \left(\frac{P_{stat}}{P_{std}} \times \frac{T_{std}}{T_{gas}} \right) \times \left(1 - B_{H_2O} \right) \tag{13.3}$$

The units for $C_{s(act)}$ are grams per actual cubic meter, or pounds per actual cubic foot.

Concentration Corrected to 50% Excess Air

Because at one time many boilers operated at 50% excess air (50% EA), this level of the air-to-fuel ratio was selected as a standard to which some pollutant concentrations would be corrected. 50% EA can also be stated as 150% of the theoretical or stoichiometric amount of air. Similarly, the *actual* amount of air mixed with the fuel is the measured %EA plus 100%. There are three versions of the equation to correct C_s to 50% EA concentration (Equations 13.4–13.6).

$$C_{s(50\%EA)} = C_s \times \left(\frac{100 + \%EA}{150}\right) \quad (13.4)$$

$$C_{s(50\%EA)} = C_s \times \left(\frac{21}{21 - 1.5[\%O_2] - 0.75[\%CO] - 0.133[\%N_2]}\right) \quad (13.5)$$

$$C_{s(50\%EA)} = C_s \times \frac{B_{CO_2} + 1.75}{2.88 - 13.9 * B_{O_2}} \quad (13.6)$$

In Equation 13.5, the percent values are measured exhaust gas percents; while in Equation 13.6, B_{O_2}, and B_{CO_2} are the mole fractions of oxygen and carbon dioxide, respectively, in the exhaust gas. All three versions give very similar results.

Concentration Corrected to 12% CO_2

A fairly common standard of dilution air for concentration corrections is 12% CO_2 in the exhaust gas (Equation 13.7).

$$C_{s(12\%CO_2)} = C_s \times \left(\frac{12}{\%CO_2}\right) \quad (13.7)$$

where percent CO_2 is the measured (EPA Method 3) value.

Other concentration corrections (e.g., 6% O_2) are similarly calculated, but infrequently used.

13.5 POLLUTANT MASS EMISSION RATE

The product of the emission test is the pollutant concentration, C_s, and, from Method 2, the gas volume flow rate:

$$Q_{s(dry,std)} = 3600 \times \left(1 - B_{H_2O}\right) \times v_s \times A \times \left(\frac{P_{stat}}{P_{std}}\right) \times \left(\frac{T_{std}}{T_{s(avg)}}\right) \quad (10.6)$$

The product of concentration and volume flow rate is the pollutant mass emission rate:

$$\frac{dm_p}{dt} = \dot{m}_p = C_s \times Q_{s(dry,std)} \tag{13.8}$$

where the pollutant mass rate has units of g/sec, g/hr, lbm/hr, kg/yr, tons/yr or, in some cases, lbm/8 hr.

The value for $d(m_p)/dt$ can be directly compared to a regulation which limits emissions per unit time; for example, the Federal New Source Performance Standard that specifies a maximum of 4.1 g/hr from a catalyst-equipped wood stove.[2]

13.6 EMISSION RATE PER UNIT OF SOME PROCESS PARAMETER

A great number of state and federal emission regulations specify the maximum allowable mass of pollutant per some measure of production or process rate. In Table 13.1, items 1, 3, 5, 6, 7, and 8 are such regulatory limits. To compare emission test results to such a regulation, the pollutant mass rate is calculated as in Section 13.5, then divided by the appropriate production rate parameter (Equation 13.9).

$$ER = \frac{\dot{m}_p}{PR} = \frac{C_s \times Q_{s(dry,std)}}{PR} \tag{13.9}$$

where ER is the emission rate and PR is the process rate.

Example Problem 13.1. Emission Compliance Check

A Portland cement kiln has a solid feed rate of 1000 tons/hr during a particulate emissions test. Test results were

C_s = 0.019 lb/dscf

Q_s = 20,000 dscfh

Based on the regulation in Table 13.1, is the source in compliance with the emission standard?

Solution

Using Equation 13.4, ER = $C_s \times Q_s/PR$ = 0.019 × 20,000/1000 = 0.38 lb particulate/ton of solids. The standard is 0.3; thus, the source is emitting more than the allowed amount of particulate.

13.7 EMISSION RATE FOR FUEL BURNING EQUIPMENT

For fuel burning source of pollutants, the process rate parameter (PR) is the energy input rate or firing rate, Q_H, in KJ/hr or million Btu per hour (mBtu/hr). The emission rate (ER) is then

$$ER = \frac{C_s \times Q_{s(dry,std)}}{Q_H} \qquad (13.10)$$

The units of ER are generally grams of pollutant per kiloJoule or pounds of pollutant per million Btu. There are four possible ways to obtain the value for the energy input rate, Q_H.

1. Fuel Input Rate

The easiest is from the product of the average fuel input rate during the emission test and the heating value of the fuel. The drawback of this simple approach is the often-encountered inability to check the accuracy of the fuel use rate meter. The other three approaches use other measurements to calculate or estimate Q_H and are applicable in the various situations noted.

2. Steam Production

For steam generation units, the steam production rate is usually accurately monitored. The steam production rate, divided by an accurate value for efficiency of conversion of fuel energy to steam energy gives Q_H.

Example Problem 13.2. Calculating energy input rate from steam production

A steam generating unit produces 119,000 lb/hr of dry, saturated steam at 650 °F. The water returned to the unit is at 225 °F. The unit's rated efficiency is $\eta = 78\%$. Compute the energy consumption rate.

Solution

Steam tables give for dry saturated steam at 650 °F, an enthalpy value of $h_g = 1125.2$ Btu/lb. For water at 225 °F, $h_w = 193$ Btu/lb. Therefore steam energy production rate is given by

$$\dot{Q}_{st} = \dot{m}\,\Delta h = 119{,}000 \times (1125.2 - 193) = 1.11 \times 10^8 \text{ Btu/h}$$

and

$$Q_H = \frac{\dot{Q}_{st}}{\eta} = 1.11 \times 10^8/0.78 = 142 \text{ mBtu/h}$$

3. Fuel Analysis and Exhaust Gas Flow Rate

This method is a good deal more complicated than the previous two and requires more information to compute Q_H. The "ultimate analysis" of the fuel must be known. The ultimate analysis is the determination of the exact chemical composition of the fuel without consideration of the physical form of the compounds. Such an analysis can only be performed by a chemistry lab with sophisticated equipment. The analysis is generally given in terms of the percent by weight of carbon, hydrogen, nitrogen, oxygen, and sulfur in a sample of the fuel, plus the percent of noncombustible ash and the higher heating value (HHV).

If the ultimate analysis of the fuel is available, the weight percents can be used to balance the theoretical or stoichiometric combustion equation for a given mass (e.g., 100 lb) of fuel. The stoichiometric balancing process assumes no oxygen and no unburned or partially combusted compounds in the exhaust gas. Next, the exhaust gas analysis data from Method 3 and the techniques of Section 10.5 can be used to compute the percent of excess air (%EA) actually being used in the combustion process. Since the stoichiometric equation gives the theoretical number of moles of each product produced when a given mass of fuel is burned, these values, together with the measured %EA, can be used to find the actual number of moles, n_{mf}, produced per unit mass of fuel (excluding H_2O, since the dry basis is desired).

Standard temperature and pressure and the total number of moles of products (less H_2O) found above can be input into the Ideal Gas Law to find the dry exhaust gas volume per unit mass of fuel, V_{mf}, at standard temperature (20 °C, 68 °F) and pressure, produced for combustion of a unit mass of fuel.

$$\text{dry std exhaust gas volume} = V_{mf} = \frac{n_{mf} \times R_u \times T_{std}}{P_{std}} \tag{13.11}$$

Thus, both the dry volume of exhaust gas per unit mass of fuel and the energy per unit mass (HHV) are known. Method 2 measurement gives the volume flow rate of dry exhaust gas, $Q_{s(dry,std)}$. This value, divided by the dry standard volume of gas per unit mass of fuel (V_{mf}) and multiplied by the energy per unit mass of the fuel, (HHV) gives the energy input per time (Q_H).

$$Q_H = \frac{Q_s(\text{dry,std})}{V_{mf}} \times \text{HHV} \tag{13.12}$$

Example Problem 13.3. Calculating Q_H from Fuel Analysis and Q_s

Find the energy input rate, Q_H, and particulate emission rate, ER, for coal with the following ultimate analysis:

68.3% C
5.1% H
8.3% O
0.9% N
2.5% S
14.9% ash
12,000 Btu/lb

and with these emission test data:

$Q_{s(dry,std)}$ = 3,000,000 dscfh
C_s(particulate) = 4.2×10^{-5} lb/dscf
%CO_2 in dry exhaust gas = 8.9%
%O_2 = 10.7%
%N_2 = 80.3%
%CO = 0%
B_{H_2O} = 3.8%

Solution

Balance an equation for complete combustion of the elements in the fuel at the stoichiometric air-to-fuel ratio. In 100 lb coal are the following number of moles:

n_c = 68.3/12 = 5.69 lbmoles carbon
(n_{H_2}) = 5.1/2 = 2.55 lbmoles hydrogen (H_2)
n_S = 2.5/32 = 0.08 lbmoles sulfur
n_{O_2} = 8.3/32 = 0.26 lbmoles oxygen (O_2)
n_{N_2} = 0.9/28 = 0.03 lbmoles nitrogen (N_2)

$$5.69C + 2.55H_2 + 0.08S + 0.03N_2 + 0.26O_2$$

$$+ \frac{a\,O_2 + (3.76 \times a \times N_2)}{(\text{air})} \rightarrow bCO_2 + cH_2O + dSO_2 + eN_2$$

The coefficients on the right-hand side are:

b = 5.69
c = 2.55
d = 0.08

An oxygen balance yields $(2 \times 0.26) + (2 \times a) = 2.55 + (2 \times 5.69) + (2 \times 0.08)$, giving a = 6.79, and a nitrogen balance gives: $(3.76 \times a) = 25.51$, so e = 25.54. From this balanced equation, the stoichiometric air-to-fuel ratio is found to be: $(A/F)_{stoich} = (32.3 \times 28.96)/85.1 = 10.99$. (Recall that the A/F = $\Sigma n_{air} \times M_{air}/\Sigma n_{fuel} M_{fuel}$.) The next step is to use the exhaust gas analysis to find %EA. This is done by balancing the equation for 100 moles of dry products as in Section 10.5.

$$f \times C_{5.69}H_{5.1}N_{0.06}S_{0.08}O_{0.52} + (g \times O_2) + (3.76g \times N_2) \rightarrow (8.9 \times CO_2)$$

$$+ (10.7 \times O_2) + (80.3 \times N_2) + (h \times H_2O) + (i \times SO_2)$$

Balancing

C: $f = 8.9/5.69 = 1.56$

N: $3.76g + 1.56 \times 0.06 = 80.3; \quad g = 21.33$

H: $1.56 \times 5.1 = 2 \times h; \quad h = 3.98$

S: $1.56 \times 0.08 = i; \quad i = 0.125$

$(A/F)_{act,\ moles} = [21.33 + 3.76 \times 21.33)]/f$

$= 101.53/1.56$

$= 65.08$

Then, $(A/F)_{act,\ mass} = 65.08 \times 28.96/(\Sigma M \times n = 85.1) = 22.15$

$$\%EA = \left[\frac{(A/F)_{act}}{(A/F)_{stoich}} - 1\right] \times 100 = \left[\frac{22.15}{10.99} - 1\right] \times 100 = 101.5\%$$

For 100 lbm of coal at 101.5% excess air, the total moles of dry exhaust products are:

CO_2 5.69 lbmoles

SO_2 0.08

O_2 $6.79 \times 1.015 = 6.89$

N_2 $25.54 + 25.51 \times 1.015 = 51.43$

TOTAL = 64.09 moles of products per 100 lbm coal

From the Ideal Gas Law:

$$V_{(std)100\ lb\ coal} = \frac{n \times R_u \times T_{std}}{P_{std}}$$

$$= \frac{64.09 \times 0.7302 \times 528}{1\ atm}$$

$$= 24{,}709.6\ dscf/100\ lbm\ coal$$

Using Equation 13.12 and multiplying by 100 for consistent units,

$Q_H = Q_s/(V_{per\ 100\ lbm\ coal}) \times HHV \times 100\ lbm$

$$Q_H = \frac{3 \times 10^6}{24{,}709.6} \times 12{,}000 \times 100$$

$$Q_H = 145{,}692{,}362\ Btu/hr = 145.69\ mBtu/hr$$

4. The F-factor Approach

The technique of Example Problem 3 is arduous, but it does provide some insight into the final technique for finding the emission rate per unit of energy input. The "F-factor" approach or technique is documented in 40CFR60, Appendix A, Method 19.[2] The development of the F-factor begins with restating Equation 13.12 as Equation 13.13.

$$\frac{Q_{s(dry,std)}}{Q_H} = \frac{V_{mf}}{HHV} = \frac{\dfrac{\text{volume exhaust gas}}{\text{mass fuel}}}{\dfrac{\text{energy}}{\text{mass fuel}}} = \frac{\text{volume exhaust gas}}{\text{unit fuel energy}} \qquad (13.13)$$

Thus, V_{mf}/HHV represents volume of dry exhaust gas per unit energy or heat, with units of dscm/KJ or dscf/mBtu. The F-factor is essentially a theoretical value for this ratio, which allows the involved calculations of Example Problem 3 to be avoided.

Notice that the left-hand side of Equation 13.13 is $Q_{s(dry,std)}/Q_H$ which is exactly what is needed to compute emission rate (Equation 13.10). There are a number of variants of the F-factor, the most important of which is F_d, the dry basis factor which gives the theoretical, or stoichiometric, volume of *dry* exhaust gas at standard temperature and pressure, $V_{mf,theo}$, per unit mass for the type of fuel, divided by the typical higher heating value of such a fuel.

$$F_d = \frac{V_{mf,theo}}{HHV} \qquad (13.14)$$

The only difference between the right-hand sides of Equations 13.13 and 13.14 is that in the former equation, V_{mf} is the actual dry exhaust volume per mass of fuel found from Method 3 data, while $V_{mf,theo}$ in Equation 13.14 is found from the balanced stoichiometric combustion equation for the same fuel. The two volumes are related by the excess air actually used in combustion.

$$V_{mf} = V_{mf,theo} \times (\text{excess air correction}) \qquad (13.15)$$

The excess air correction normally used assumes the oxygen content of air is 20.9% by volume and that there is negligible nitrogen in the fuel and negligible carbon monoxide in the exhaust gas. With these assumptions, Equation 10.9 for %EA becomes

$$\%EA = \frac{\%O_2}{20.9 - \%O_2} \times 100 \qquad (13.16)$$

and the excess air correction is

$$1 + \frac{\%EA}{100} = \frac{20.9}{20.9 - \%O_2} \quad (13.17)$$

Thus, Equation 13.15 becomes:

$$V_{mf} = V_{mf,theo} \times \left(\frac{20.9}{20.9 - \%O_2}\right) \quad (13.18)$$

and Equations 13.13 and 13.14 give:

$$\frac{Q_{s(dry,std)}}{Q_H} = \frac{V_{mf,theo} \times \left(\frac{20.9}{20.9 - \%O_2}\right)}{HHV} = F_d \times \left(\frac{20.9}{20.9 - \%O_2}\right) \quad (13.19)$$

Finally,

$$ER = C_s \times F_d \times \left(\frac{20.9}{20.9 - \%O_2}\right) \quad (13.20)$$

Again, F_d is a theoretical value for the volume of dry exhaust gas per unit of fuel energy at a stoichiometric air-to-fuel ratio. For each unit of mass of a fuel whose ultimate analysis is known, an easily calculated dry standard volume of gaseous products will be produced in stoichiometric combustion. In metric units,

$$V_{mf,theo} = 227.0(\%H) + 95.7(\%C) + 35.4(\%S) + 8.6(\%N) - 28.5(\%O) \quad (13.21)$$

where the percentages are from the ultimate analysis and the volume is in dscm. Dividing the value in Equation 13.21 by HHV gives F_d.

There are other forms of the F-factor for other methods of computing the excess air correction. Continuous gas emission monitoring personnel often use a CO_2-based method to find the excess air, and use F_c values:

$$ER = C_s \times F_c \times \left(\frac{100}{\%CO_2}\right) \quad (13.22)$$

Wet-basis F-factors are also sometimes used and are discussed in EPA documentation.[2]

F-factors for various typical fuels have been compiled in 40CFR60, Appendix A, Method 19[2] (see Figure 13.3).

Fuel type	F_d		F_w		F_c	
	dscm/J	dscf/10^6 Btu	wscm/J	wscf/10^6 Btu	scm/J	scf/10^6 Btu
Coal:						
Anthracite [2]	2.71×10^{-7}	10,100	2.83×10^{-7}	10,540	0.530×10^{-7}	1,970
Bituminous [2]	2.63×10^{-7}	9,780	2.86×10^{-7}	10,640	0.484×10^{-7}	1,800
Lignite	2.65×10^{-7}	9,860	3.21×10^{-7}	11,950	0.513×10^{-7}	1,910
Oil [3]	2.47×10^{-7}	9,190	2.77×10^{-7}	10,320	0.383×10^{-7}	1,420
Gas:						
Natural	2.43×10^{-7}	8,710	2.85×10^{-7}	10,610	0.287×10^{-7}	1,040
Propane	2.34×10^{-7}	8,710	2.74×10^{-7}	10,200	0.321×10^{-7}	1,190
Butane	2.34×10^{-7}	8,710	2.79×10^{-7}	10,390	0.337×10^{-7}	1,250
Wood	2.48×10^{-7}	9,240	...	...	0.492×10^{-7}	1,830
Wood Bark	2.58×10^{-7}	9,600	...	...	0.516×10^{-7}	1,920
Municipal	2.57×10^{-7}	9,570	...	...	0.488×10^{-7}	1,820
Solid Waste	...					

[1] Determined at standard conditions: 20 °C (68 °F) and 760 mm Hg (29.92 in. Hg).
[2] As classified according to ASTM D388–77.
[3] Crude, residual, or distillate.

Figure 13.3. F-factors for various fuels.

CHAPTER 13 PROBLEMS

1. Below are data from a Method 5 Test. These are the same data as in Problems 2 and 3 in Chapter 12. Calculate the emission rate in pounds per million Btu. Assume the fuel is #6 heating oil and use the F_d value from Figure 13.3.

Stack Test Data for Problem 1

V_{dgm}	37.2	Dry gas volume, actual ft^3
P_{MASS}	100.3	mg
$(\Delta P_p)^{1/2}_{avg}$	0.14	average square root of ΔP, $(''H_2O)^{1/2}$
P_{atm}	29.29	Inches Hg
P_{stat}	−0.240	Gage press, " H_2O
DIA	0.5	Stack diameter, ft
C_{dgm}	0.935	Y_{avg}, the dry gas meter coefficient
C_P	0.85	Pitot coefficient
T_{dgm}	538	Temperature of gas in DGM, °R
V_C	5.4	Water condensed, mL
m_{sg}	17.3	Mass adsorbed onto silica gel, g
ΔH_{avg}	0.97	Average orifice head loss
CO_2	7.0	%
O_2	15.9	%
CO	0.0	%
T_S	717	°R
D_n	0.5	Nozzle diameter, inch (Area, $A_n = \pi D_n^2/4$ = 0.1963/144 = (0.00136 ft^2)
FR	5500	Fuel consumption rate, lbm/hr
θ	60	Total test time, min
HHV	150000	Energy content of fuel, Btu/lbm
T_{std}	528	°R
P_{std}	29.92	Inches Hg

2. An emissions monitoring company ran a CEM unit plus EPA Methods 2 and 4 on the stack of a small coal furnace. The data measured follow:

   v_s = 14 m/sec
   Stack diameter = 0.24 m
   NO_x concentration in the exhaust gas = 765 ppm
   B_{H_2O} = 4.7%
   Stack pressure = 783 mm Hg
   Stack temperature = 455 °C
   % CO_2 in exhaust gas = 10%

Calculate the NO_x emission rate in grams per second (g/sec).

APPENDIX A
CONVERSION TABLES

QUANTITIES OF UNITS COMMONLY FOUND IN THE LITERATURE ON AIR POLLUTION

1 acre	1/640 mi^2	$4.047 \times 10^3\ m^2$
1 Angstrom (Å)	10^{-8} cm	10^{-10} m
1 atmosphere (atm)	1.013×10^6 dyn cm^2	$1.013 \times 10^5\ N/m^2$
1 bar (b)	100 KPa	$10^5\ N/m^2$
1 barrel (bbl)	42 gal, U.S.	$0.159\ m^3$
1 boiler horsepower	3.35×10^4 btu/hr	9.810×10^3 W
1 British Thermal Unit (Btu)	252 cal	1.054×10^3 J
1 Btu hours	1.93×10^6 erg-sec	0.293 W
1 calorie (cal)	4.184×10^{-7} erg	4.184 J
1 centimeter of mercury (cm Hg)	$1.333 \times 10^4\ dyn/cm^2$	$1.333 \times 10^3\ N/m^2$
1 cubic foot, U.S. (cu. ft)	$2.832 \times 10^4\ cm^3$	$2.832 \times 10^{-2}\ m^3$
1 dyne (dyn)	1 $g\text{-}cm/sec^2$	10^{-5} N
1 erg	1 $g\text{-}cm^2/sec^2$	10^{-7} J
1 foot, U.S. (ft)	30.48 cm	0.3048 m
1 foot per minute (ft/min)	1.829×10^{-2}km/hr	5.080×10^{-3} m/sec
1 gallon, U.S. (gal)	$3.785 \times 10^3\ cm^3$	$3.785 \times 10^{-3}\ m^3$

CONVERSION FACTORS

Capacity, Energy, Force, Heat

Multiply	By	To obtain
Btu	0.252	Kilogram-calories
Btu	9.48×10^{-4}	Watt-seconds (Joules)
Btu	3.927×10^{-4}	Horsepower hours
Btu	2.928×10^{-4}	Kilowatt-hours
Btu/min	0.02356	Horsepower
Btu/min	0.01757	Kilowatts
Btu/min	10^{-3}	Pound-hour steam
Horsepower (boiler)	33.479	Btu/hour
Horsepower (boiler)	9.803	Kilowatts
Horsepower-hours	0.7457	Kilowatt-hours
Kilowatts	56.92	Btu/minute
Kilowatts	1.341	Horsepower
Kilowatt-hours	3415	Btu
Kilowatt-hours	1.341	Horsepower
Megawatts	1360	Kilogram-hour steam
Pound-hour steam	0.434	Kilogram-hour

Transfer Coefficient

Multiply	By	To obtain
Btu/(hr)(ft^2)(°F)	0.001355	Cal/(sec)(cm^2)(°C)
	1.929×10^{6}	Btu/(sec)(in^2)(°F)
	0.0005669	Watts/(cm^2)(°C)

Flow

Multiply	By	To obtain
Cubic feet/minute	0.1247	Gallons/second
Cubic feet/second	0.646317	Million gallons/day
Cubic feet/second	448.831	Gallons/minute
Cubic meter/second	22.8	Million gallons/day
Cubic meter/second	8.32×10^{9}	Gallons/year
Gallons/year	10.37×10^{-6}	Cubic meters/day
Gallons/minute	2.228×10^{-3}	Cubic feet/second
Liters/minute	5.886×10^{-4}	Cubic feet/second
Liters/minute	4.403×10^{-3}	Gallons/second
Million gallons/day	1.54723	Cubic feet/second
Million gallons/day	0.044	Cubic meters/second
Million gallons/day	695	Gallons/minute
Pounds of water/minute	2.679×10^{-4}	Cubic feet/second

Length, Area, Volume

Multiply	By	To obtain
Acres	43,560	Square feet
Acres	4047	Square meters
Acres	1.562×10^{-3}	Square miles
Barrels oil	0.156	Cubic meters
Barrels oil	42	Gallons oil
Centimeters	0.3937	Inches
Cubic feet	2.832×10^{4}	Cubic centimeters
Cubic feet	1728	Cubic inches
Cubic feet	0.02832	Cubic meters
Cubic feet	0.03704	Cubic yards
Cubic feet	7.48052	Gallons
Cubic feet	28.32	Liters
Cubic meters	35.31	Cubic feet
Cubic meters	264.2	Gallons

Multiply	By	To obtain
Feet	30.48	Centimeters
Feet	0.3048	Meters
Gallons	0.1337	Cubic feet
Gallons	3.785×10^{-3}	Cubic meters
Gallons	3.785	Liters
Gallons water	8.3453	Pounds water
Liters	0.2642	Gallons
Meters	3.281	Feet
Meters	39.37	Inches
Square feet	2.296×10^{-5}	Acres
Square feet	0.09290	Square meters
Square meters	2.471×10^{-4}	Acres
Square meters	10.76	Square feet
Square miles	640	Acres

Mass, Pressure, Temperature, Concentration

Multiply	By	To obtain
Atmospheres	29.92	Inches Hg
Atmospheres	33.90	Feet water
Atmospheres	14.70	Pounds/square inch
Feet water	0.02947	Atmospheres
	0.04335	Pounds/square inch
	62.378	Pounds/square foot

Mass, Pressure, Temperature, Concentration (continued)

Inches Hg	0.03342	Atmospheres
	13.60	Inches water
	1.133	Feet water
	0.4912	Pounds/square inch
	70.727	Pounds/square foot
Inches water	0.03609	Pounds/square inch
	5.1981	Pounds/square foot
Kilograms	2.2046	Pounds mass (lbm)
Pounds mass	453.5924	Grams
Pounds water	0.01602	Cubic feet
Pounds water	0.1198	Gallons
Pounds/square inch	0.06804	Atmospheres
Pounds/square inch	2.307	Feet water
Pounds/square inch	2.036	Inches Hg
Temp. (°C) + 17.78	1.8	Temperature (°F)
Temp. (°F) – 32	0.555	Temperature (°C)
Kelvin = degrees Celsius + 273.16		
Degrees Rankin = degrees Fahrenheit + 459.69		
Tons (metric)	2205	Pounds
Tons (U.S.)	2000	Pounds
Tons (short)	0.89287	Tons (long)
Tons (short)	0.9975	Tons (metric)

Thermal Conductivity

Multiply	By	To obtain
Btu/(hr)(ft^2)(°F/ft)	0.00413	Cal/(sec)(cm^2)(°C cm)
	12	Btu/(hr)(ft^2)(°F/in)

Viscosity

Multiply	By	To obtain
Poise	1.0	Gm/cm-sec
	1.0	Dyne-sec/cm^2
	100	Centipoise (C_p)
Centipoise	0.000672	Pounds/foot-second
	0.0000209	Pound/second square foot
	2.42	Pound/foot-hour
Stoke	1.0	Square centimeter/second
	0.155	Squared inch/second
	0.001076	Squared foot/second
	density (gm/cm^3)	Poise

Density

Multiply	By	To obtain
Grams per cc	62.428	Pounds/cubic foot
	0.03613	Pounds/cubic inch
	8.345	Pounds/gallon
Gram-moles of Ideal Gas at 0°C and 760 mm Hg	22.4140	Liters
Pounds per cubic inch	1728	Pounds/cubic foot
	27.68	Grams/cubic centimeter
Pounds-moles of Ideal Gas at 0°C and 760mm Hg	359.05	Cubic feet
Grams/liter	58.417	Grains/gallon
Grams/liter	8.345	Pounds/1000 gallons
Grams/liter	0.062427	Pounds/cubic foot
Parts/million (ppm by wt)	0.0584	Grains/gallons
Parts/million (ppm by wt)	8.345	Pounds/million gallons

CONVERSION CONSTANTS (SI)

Multiply	By	To obtain
Milligrams/m^3	1000	Micrograms/m^3
	1.0	Micrograms/liter
	(24.45/Mgas)	Ppm by volume (25 °C)
	62.43×10^{-9}	lbm/ft^3
Micrograms/m^3	0.001	Milligrams/m^3
	0.001	Micrograms/liter
	0.02445/Mgas	Ppm by volume (25 °C)
	62.43×10^{-12}	lbm/ft^3
Micrograms/liter	1.0	Milligrams/m^3
	1000	Micrograms/m^3
	24.45/Mgas	Ppm by volume (25 °C)
	62.43×10^{-9})	lbm/ft^3
Ppm by volume (25 °C)	(Mgas/24.45)	Milligrams/m^3
	(Mgas/0.02445)	Micrograms/m^3
	(Mgas/24.45)	Micrograms/liter
lbm/ft^3	16.018×10^6	Milligrams/m^3
	16.018×10^9	Micrograms/m^3
	16.018×10^6	Micrograms/liter
	133.7×10^3	ppm by wt

CONVERSION CONSTANTS (SI) (continued)

Grams/m^3	1000.0	Milligrams/m^3
	0.02832	Grams/ft^3
	1.0×10^6	Micrograms/m^3
	28.317×10^3	Micrograms/ft^3
	0.06243	lbm/1000 ft^3
No. of particles/ft^3	35.314	No./m^3
	35.314×10^{-3}	No./L
	35.314×10^{-6}	No./cm^3
Tons/mi^2	3.125	lbm/acre
	0.07174	lbm/1000 ft^2
	0.3503	Grams/m^2
	350.3	Milligrams/m^2
	0.03503	Milligrams/cm^2
lbm	7000.0	Grains
Micrometer	3.937×10^{-5}	in
	1.0×10^{-3}	mm

CONSTANTS AND USEFUL INFORMATION

Energy Equivalences of Various Fuels

	Approximate values
Bituminous coal	22×10^6 Btu/ton
Anthracite coal	26×10^6 Btu/ton
Lignite coal	16×10^6 Btu/ton
Residual oil	147,000 Btu/gal
Distillate oil	140,000 Btu/gal
Natural gas	1000 Btu/ft^3

1 lb of water evaporated from and at 212 °F equals:

0.2844 Kilowatt-hours
0.3814 Horsepower-hours
970.2 Btu

1 cubic ft air weights 34.11 gm.

Miscellaneous Physical Constants

Multiply	By	To obtain
Avogadro's Number	6.0228×10^{23}	Molecules/g-mole
Gas Law Constant, R	1.987	Cal/(g-mole)(K)
	1.987	Btu/(lb-mole) (°R)
	82.06	(cm^3)(atm)/(g-mole) (K)
	10.731	(ft^3)(lb)(in.^{2})/(lbmole)(°R)
	0.7302	(ft^3)(atm)/(lbmole) (°R)

APPENDIX B
TOXIC ORGANIC METHODS

METHODS FOR DETERMINATION OF TOXIC ORGANIC COMPOUNDS IN AIR (FROM REFERENCE 11)

TO1 Adsorption on Tenax® and analysis by GC/MS; for volatile non-polar organics with boiling points from 80 to 200 °C.

TO2 Adsorption on carbosieve and analysis by GC/MS; for highly volatile nonpolar organics with boiling points from –15 to 120 °C.

TO3 Cryogenic trapping and analysis by GC/FID or ECD; for volatile non-polar organics with boiling points from –10 to 200 °C.

TO4 High volume PUF sampling and analysis by GC/ECD; for organochlorine pesticides and PCBs.

TO5 Dinitrophenylhydrazine absorption and analysis by HPLC/UV; for aldehydes and ketones.

TO6 Absorption in aniline/toluene and analysis by HPLC; for phosgene.

TO7 Adsorption on Thermosorb/N and solvent desorption with GC/MS analysis; for *N*-nitrosodimethylamine.

TO8 Absorption in sodium hydroxide and analysis by HPLC; for cresol and phenol.

TO9 High volume PUF sampling and analysis by high-resolution GC/MS; for dioxin.

TO10 Low volume PUF sampling and analysis by GC/ECD; for pesticides.

TO11 Adsorption on silica gel coated with dinitrophenylhydrazine, solvent desorption and analysis by HPLC; for formaldehyde.

TO12 Cryogenic preconcentration and direct FID analysis; for non-methane organic compounds.

TO13 Collection/adsorption on PUF followed by XAD-2 and analysis by solvent desorption and HPLC; for polynuclear aromatic hydrocarbons.

TO14 Whole air sampling with SUMMA® passivated canister and analysis by GC and various detectors; for semivolatile and volatile organic compounds.

APPENDIX C EMISSION MEASUREMENT METHODS

THE EMISSION TEST METHODS LISTED HERE ARE FULLY DESCRIBED IN 40CFR60, APPENDIX A, EXCEPT AS NOTED.

Method 1	Sample and velocity traverses for stationary sources.
Method 1A	Sample and velocity traverses for stationary sources with small stack or ducts.
Method 2	Determination of stack gas velocity and volumetric flow rate (type S pitot).
Method 2A	Direct measurement of gas volume through pipes and small ducts.
Method 2B	Determination of exhaust gas volume flow rate from gasoline vapor incinerators.
Method 2C	Determination of stack gas velocity and volumetric flow rate in small stacks or ducts (standard pitot tube).
Method 2D	Measurement of gas volumetric flow rates in small pipes and ducts.
Method 3	Gas analysis for CO, O_2, excess air, and dry molecular weight.
Method 3A	Determination of O_2 and CO_2 concentrations in emissions from stationary sources (instrumental analyzer procedure).
Method 4	Determination of moisture content in stack gases.
Method 5	Determination of particulate emissions from stationary sources.

Method 5A	Determination of particulate emissions from the asphalt processing and asphalt roofing industry.
Method 5B	Determination of nonsulfuric acid particulate matter from stationary sources.
Method 5C	(Reserved.)
Method 5D	Determination of particulate emissions from positive pressure fabric filters.
Method 5E	Determination of particulate emissions from the wool fiberglass insulation manufacturing industry.
Method 5F	Determination of nonsulfate particulate matter from stationary sources.
Method 5G	Determination of particulate emissions from wood heaters from a dilution tunnel sampling location.
Method 5H	Determination of particulate emissions from wood heaters from a stack location.
Method 6	Determination of sulfur dioxide emissions from stationary sources.
Method 6A	Determination of sulfur dioxide, moisture, and carbon dioxide emissions from fossil fuel combustion sources.
Method 6B	Determination of sulfur dioxide and carbon dioxide daily aver age emissions from fossil fuel combustion sources.
Method 6C	Determination of sulfur dioxide emissions from stationary sources (instrumental analyzer procedure).
Method 7	Determination of nitrogen oxide emissions from stationary sources.
Method 7A	Determination of nitrogen oxide emissions from stationary sources (ion chromatographic method).
Method 7B	Determination of nitrogen oxide emissions from stationary sources (ultraviolet spectrophotometry).
Method 7C	Determination of nitrogen oxide emissions from stationary sources (alkaline-permanganate/colorimetric method).
Method 7D	Determination of nitrogen oxide emissions from stationary sources (alkaline-permanganate/ion chromatographic method).
Method 7E	Determination of nitrogen oxide emissions from stationary sources (instrumental analyzer procedure).
Method 8	Determination of sulfuric acid mist and sulfur dioxide emissions from stationary sources.
Method 9	Visual determination of the opacity of emissions from stationary sources.
Method 10	Determination of carbon monoxide emissions from stationary sources.
Method 10A	Determination of carbon monoxide emissions in certifying continuous emission monitoring systems at petroleum refineries.

Method 10B	Determination of carbon monoxide emissions from stationary sources.
Method 11	Determination of hydrogen sulfide content of fuel gas streams in petroleum refineries.
Method 12	Determination of inorganic lead emissions from stationary sources.
Method 13A	Determination of total fluoride emissions from stationary sources (SPADNS zirconium lake method).
Method 13B	Determination of total fluoride emissions from stationary sources (specific ion electrode method).
Method 14	Determination of fluoride emissions from potroom roof monitors for primary aluminum plants.
Method 15	Determination of hydrogen sulfide, carbonyl sulfide, and carbon disulfide emissions from stationary sources.
Method 15A	Determination of total reduced sulfur emissions from sulfur recovery plants in petroleum refineries.
Method 16	Semicontinuous determination of sulfur emissions from stationary sources.
Method 16A	Determination of total reduced sulfur emissions from stationary sources (impinger technique).
Method 16B	Determination of total reduced sulfur emissions from stationary sources.
Method 17	Determination of particulate emissions from stationary sources (in-stack filtration method).
Method 18	Measurement of gaseous organic compound emissions by gas chromatography.
Method 19	Determination of sulfur dioxide removalefficiency and particulate, sulfur dioxide, and nitrogen oxides emission rates.
Method 20	Determination of nitrogen oxides, sulfur dioxide, and diluent emissions from stationary gas turbines.
Method 21	Determination of volatile organic compound leaks.
Method 22	Visual determination of fugitive emissions from material sources and smoke emissions from flares.
Method 23	Determination of polychlorinated dibenzo-*p*-dioxins and polychlorinated dibenzofurans from stationary sources.
Method 24	Determination of volatile matter content, water content, density, volume of solids, and weight of solids of surface coatings.
Method 24A	Determination of volatile matter content and density of printing inks and related coatings.
Method 25	Determination of total gaseous nonmethane organic emissions as carbon.
Method 25A	Determination of total gaseous organic concentration using a flame ionization analyzer.

Method 25B	Determination of total gaseous organic concentration using a nondispersive infrared analyzer.
Method 26	Determination of hydrogen chloride emissions from stationary sources.
Method 27	Determination of vapor tightness of gasoline delivery tank using pressure-vacuum test.
Method 28	Certification and auditing of wood heaters.
Method 28A	Measurement of air-to-fuel ratio and minimum achievable burn rates for wood-fired appliances.

FROM EPA TEST METHODS FOR EVALUATING SOLID WASTE, SW-846:

Method 0010	Modified Method 5 sampling for the determination of destruction and removal efficiency of semivolatile principal organic hazardous compounds from incineration systems.
Method 0020	Source assessment sampling for semiquantitative estimates of the amounts and types of semivolatile organic and particulate materials discharged from incinerators.
Method 0030	Volatile organic sampling for the determination of volatile principal organic hazardous compounds in the effluent of hazardous waste incinerators.

APPENDIX D METHOD 5 FORTRAN PROGRAM

FORTRAN PROGRAM FOR METHOD 5 CALCULATIONS

```
      PROGRAM MAIN
C     PROGRAM FOR STACK CALCULATIONS, EPA METHOD 5
C2345678
      CHARACTER*21 SORCE*31,DATE,DATOUT*80,DATIN*80
      REAL
     1 N2,TSTD,PSTD,SO2,VM,PMASS,SQD,PBAR,PS,HT,WD,CDGM,CP,
     2 TM,VC,SGM,OP,CO2,O2,C0,HHV,TS,DIAN,FR,TIME,VMSTD,MWD,
     3 MWT,EA,BH2O,AREA,AN,XISO
C     this program does not deal with SO2, tho other versions might
      SO2 = 0.0
C     standard temperature and pressure (note that the program uses English
      units throughout)
      TSTD = 528.
      PSTD = 29.92
C     ****** SET UP AND OPEN OUTPUT FILE **********
      WRITE(6,97)
97    FORMAT(1X,'ENTER INPUT FILENAME')
      READ(5,100) DATIN
      OPEN(UNIT=2,FILE=DATIN(1:80),STATUS ='OLD')
      WRITE(6,99)
99    FORMAT(1X,'ENTER OUTPUT FILENAME')
      READ(5,100) DATOUT
100   FORMAT(A80)
      OPEN(UNIT=3,FILE=DATOUT(1:80),STATUS ='NEW')
C     first data item is identifier for unit tested, 31 ALPHA CHARACTERS
```

```
      READ(2,*)SORCE
C     next input is date, up to 21 ALPHA CHARACTERS
      READ(2,*)DATE
C     input DGM total reading in cubic feet
      READ(2,9) VM
9     FORMAT(15X,F10.0)
C     input mass of particulate collected in milligrams
      READ(2,9)PMASS
C     input avg of square root of pitot delta p
      READ(2,9)SQD
C     input barometric pressure in inches Hg
      READ(2,9)PBAR
C     input stack gage static pressure in inches of H2O
      READ(2,9)PS
C TO CHANGE TO ABSOLUTE PRESSURE
      PS1 = PBAR + PS/13.6
C     input stack diameter, if circular, in inches; 0 if rectangular
      READ(2,9)DIA
C     input  DGM cal factor, CDGM
      READ(2,9)CDGM
C     input the pitot cal factor CP
      READ(2,9)CP
C     input the avg DGM temp in degrees RANKINE
      READ(2,9)TM
C     input the volume of water collected in impingers, milliliters
      READ(2,9)VC
C     input the mass change of the silica gel impinger in grams
      READ(2,9)SGM
C     input the average orifice pressure drop, delta H in inches H2O
      READ(2,9)OP
C     input the orsat data in order:  %CO2, %O2, %CO
      READ(2,9)CO2
      READ(2,9)O2
      READ(2,9)CO
C     input moisture content of fuel in % (probably zero, but must enter)
      READ(2,9)DMOIS
C     input average stack gas temp in degrees RANKINE
      READ(2,9)TS
C     input nozzle diameter in inches
      READ(2,9)DIAN
C     input the fuel use rate, if known, in gal/hr, otherwise zero
      READ(2,9)FR
C     input total sample time in minutes
      READ(2,9) TIME
C     input the fuel heating value in Btu/gal
      READ(2,9)HHV
C     input rectangular duct width and height in feet (use 0 if circular)
      READ(2,9)WD
```

```
      READ(2,9)HT
      ENGR = FR*(HHV/1000000.)
      WRITE(3,240)
240   FORMAT(20X,'INPUT DATA '///)
264   WRITE(3,270)VM
270   FORMAT(15X,'VM = ',F10.3,' FT3 (VOLUME OF DRY GAS METER)')
      WRITE(3,280)PMASS
280   FORMAT(15X,'PMASS = ',E15.8,' MG (MASS OF PARTCULATE)')
      WRITE(3,290)SQD
290   FORMAT(15X,'SQD = ',F10.3,' IN. H20 (AVG. SQ. RT OF PITOT PRESS)')
      WRITE(3,300)PBAR
300   FORMAT(15X,'PBAR = ',F10.3,' IN. HG (BAROMETRIC PRESSURE)')
      WRITE(3,310)PS
310   FORMAT(15X,'PS = ',F10.3,' IN. H2O (STATIC PRESSURE)')
      WRITE(3,320)DIA
320   FORMAT(15X,'DIA = ',F10.3,' INCH (STACK DIAMETER - 0 FOR RECT)')
C     WRITE(3,321)HT,WD
321   FORMAT(15X, 'STACK DIMENSIONS',F10.3,' X ',F10.3,' FT')
      WRITE(3,350)PSTD
350   FORMAT(15X,'PSTD = ',F10.3,' IN. HG (STD PRESSURE)')
      WRITE(3,360)TSTD
360   FORMAT(15X,'TSTD = ',F10.3,' R (STD TEMPERATURE)')
      WRITE(3,370)TM
370   FORMAT(15X,'TM = ',F10.3,' R (AVG DRY GAS METER TEMP)')
      WRITE(3,380)VC
380   FORMAT(15X,'VC = ',F10.3,' ML (VOLUME OF MOISTURE COLLECTED)')
      WRITE(3,390)SGM
390   FORMAT(15X,'SGM = ',F10.3,' GM (MASS OF SILICA GEL COLLECTED)')
      WRITE(3,400)CO2
400   FORMAT(15X,'CO2 = ',F10.3, '(CARBON DIOXIDE)')
      WRITE(3,410)O2
410   FORMAT(15X,'O2 = ',F10.3, '(OXYGEN)')
      WRITE(3,420)CO
420   FORMAT(15X,'CO = ',F10.3, '( CARBON MONOXIDE)')
      WRITE(3,440)TS
440   FORMAT(15X,'TS = ',F10.3,' R (AVG STACK TEMPERATURE)')
C     WRITE(3,442) DMOIS
442   FORMAT(15X,'DMOIS = ',F10.3,' % (FUEL MOISTURE,DRY BASIS)')
      WRITE(3,445)OP
445   FORMAT(15X,'OP = ',F10.3,' IN. H20 (DELTA H)')
C     WRITE(3,446)FR
446   FORMAT(15X,'FR = ',F10.3,' GAL/HR (FUEL RATE) ')
      WRITE(3,447)DIAN
447   FORMAT(15X,'DIAN = ',F10.4,' INCH (NOZZLE DIAMETER)')
      WRITE(3,448)CDGM
448   FORMAT(15X,'Y =   ',F10.3,'(CAL FACTOR FOR DRY GAS METER)')
      WRITE(3,450)TIME
450   FORMAT(15X,'TIME = ',F10.3,'(TOT MINUTES OF SAMPLING)'////)
```

```
C       WRITE(3,449)HHV
449     FORMAT(15X,'HHV = ',F10.3,'(ENERGY OF THE FUEL, BTU/GAL)'////)
        VMSTD = (((((PBAR+(OP/13.6))/PSTD)*(TSTD/TM)))*CDGM)*VM
        N2 = 100.-CO2-O2-CO
C       AN is nozzle area in square feet
        AN = (((DIAN/12.)**2)*3.14159)/4.0
        AREA = HT*WD
        IF (AREA.EQ. 0.0)THEN
C       AREA is stack cross sectional area in ft2
        AREA = 3.14159*(DIA**2)/4.0/144
        ENDIF
C       IDENTITY OF VARIABLES
C       HHV = higher heating value of fuel
C       MWD = DRY MOLEC WT
C       MWT = TOTAL MOLEC WT
C       N2 = % NITROGEN
C       ER1 = EMISSION RATE IN LB/MBTU
C       XISO = % ISOKINETIC
C       AN = NOZZLE AREA IN FEET
C       ER = EMISSION RATE IN LB/HR
C       VS = STACK GAS VELOCITY IN FT/SEC
C       QS = STACK GAS VOL FLOWRATE IN DSCFH
C       CS = CONCENTRATION OF PARTICULATE IN EXHAUST, LB/FT3
C       V1C = TOTAL WATER COLLECTED = VC + SGM in ML
        CALL MOLWT(CO2,O2,CO,VC,SGM,N2,BH2O,MWD,MWT,EA,
     1 VMSTD,V1C,VWSTD)
        CALL EMISRT(PMASS,TS,PS1,CP,MWT,VMSTD,SQD,VS,QS,CS,
     1 BH2O,PSTD,TSTD,AREA,FR,HHV,ER1,ER,O2)
        CALL PCTISO(TS,PSTD,VMSTD,TSTD,PS1,BH2O,VS,TIME,AN,XISO)
        CALL OUTFILE(SORCE,DATE,VMSTD,VWSTD,V1C,BH2O,MWD,
     1 MWT,CS,EA,VS,QS,ER1,XISO,ER)
C ++++++++++++++++++++ END OF MAIN ++++++++++++++++++++
C ---------------------------------------------------------
        STOP
        END
        SUBROUTINE MOLWT(CO2,O2,CO,VC,SGM,N2,BH2O,MWD,MWT,
     1 EA,VMSTD,V1C,VWSTD)
C called by STACK.FOR for method 5 calculations
        REAL CO2,O2,CO,VC,SGM,N2,BH2O,MWD,MWT,EA,V1C,VWSTD,
     1 VMSTD
C VOLUME OF GAS AT STP, IN FT3
        V1C = VC + SGM
C TOTAL MOISTURE COLLECTED IN ML
        VWSTD = (0.0472 * V1C)
C TOTAL WATER VAPOR, IN FT3
        BH2O=VWSTD / (VMSTD + VWSTD)
C MOISTURE CONTENT OF STACK, NO UNITS
        MWD = ((O2*32.)+(CO2*44.)+(CO*28.)+(N2*28.))/100.
```

```
C DRY MOLECULAR WEIGHT
      MWT=(MWD*(1.-BH2O))+(18.*BH2O)
C TOTAL MOLECULAR WEIGHT, LB/LB-MOLE
      EA=(((O2-(.5*CO)))/(((.264*N2)-O2)+(.5*CO)))*100.
C    EA is % EXCESS AIR
      RETURN
      END
C **********************************
      SUBROUTINE EMISRT(PMASS,TS,PS1,CP,MWT,VMSTD,SQD,VS,QS,
     1 CS, BH2O,PSTD,TSTD,AREA,FR,HHV,ER1,ER,O2)
      REAL CS,VS,QS,ER,ER1,MWT
      CS = PMASS/VMSTD
      VS = 85.49*CP*(SQRT(TS*(PS1*MWT)))*SQD
      QS = (AREA*VS)*(1-BH2O)*(PSTD/PS1)*(TS/TSTD)*3600
      IF(FR.EQ.1)THEN
300   PRINT*,'ENTER YOUR DESIRED FUEL TYPE:'
      PRINT*,'ENTER (1) TO QUIT:'
      PRINT*,'BITUMINOUS COAL - 2'
      PRINT*,'RESIDUAL OIL - 3'
      PRINT*,'NATURAL GAS - 4'
      PRINT*
      READ*,FR
      IF(FR.EQ.2)THEN
      FR=9820
      ELSE IF(FR.EQ.3)THEN
      FR=9220
      ELSE IF(FR.EQ.4)THEN
      FR=8740
      ELSE
      PRINT*,'ALLOWABLE ENTRIES INCLUDE 1,2,3,4:'
      ENDIF
      ENDIF
      ER=FR*CS*(20.9/((20.9*(1-BH2O))-(O2/(1-BH2O))))
      ER1 =  QS*ER
      RETURN
      END
C **********************************
      SUBROUTINE PCTISO(TS,PSTD,VMSTD,TSTD,PS1,BH2O,VS,TIME,
     1 AN,XISO)
      REAL TS,PSTD,VMSTD,TSTD,PS1,BH2O,VS,TIME,AN,XISO
C XISO IS PERCENT ISOKINETIC SAMPLING
      XISO = (TS*PSTD*VMSTD)(((TSTD*PS1)*(1.BH2O))*(VS*60.*TIME*
     1 AN))*100.
      RETURN
      END
C     SUBROUTINE OUTFILE(SORCE,DATE,VMSTD,VWSTD,V1C,BH2O,
     1 MWD,MWT,CS,EA,VS,QS,ER1,XISO,ER)
      REAL VMSTD,VWSTD,V1C,BH2O,MWD,MWT,CS,EA,VS,QS,ER1,XISO
```

```
      CHARACTER*21 SORCE*31,DATE
      WRITE(3,*)'THE UNIT TESTED WAS ',SORCE
      WRITE(3,*)'THE DATE OF TESTING WAS ',DATE
90    WRITE(3,100)VMSTD
100   FORMAT(15X,'VOLUME GAS COLLECTED AT STP =',2X,F10.5,1X,
    1 FT3'/)
      WRITE(3,110)VWSTD
110   FORMAT(15X,'VOLUME VAPOR AT STP=',2X,F10.5,1X,'FT3'/)
      WRITE(3,120)V1C
120   FORMAT(15X,'TOTAL MOISTURE COLLECTED =',2X,F10.5,1X,'G'/)
      WRITE(3,130)BH2O
130   FORMAT(15X,'WATER VAPOR FRACTION =',2X,F10.5/)
      WRITE(3,140)MWD
140   FORMAT(15X,'DRY MOLECULAR WEIGHT =',2X,F10.5,1X,'LB/LB
    1 MOLE'/)
      WRITE(3,150)MWT
150   FORMAT(15X,'TOTAL MOLECULAR WEIGHT =',2X,F10.5,1X,'LB/LB
    1 MOLE'/)
      WRITE(3,160)CS
160   FORMAT(15X,'PARTICULATE CONCENTRATION =',2X,E15.7,1X,
    1 'LB/DSCF'/)
      WRITE(3,170)EA
170   FORMAT(15X,'% EXCESS AIR =',2X,G15.5,1X,'%'/)
      WRITE(3,180)VS
180   FORMAT(15X,'STACK GAS VELOCITY =',2X,F10.5,1X,'FT/SEC'/)
      WRITE(3,190)QS
190   FORMAT(15X,'STACK GAS FLOW RATE =',2X,G15.5,1X,'DSCFH'/)
      WRITE(3,200)ER
200   FORMAT(15X,'EMISSION RATE =',2X,G15.5,1X,'LB/HR'/)
      ER1 = CS*7000
C     FOR THIS VERSION, ER1 IS CHANGED TO GRAINS/DSCF
      WRITE(3,210)ER1
210   FORMAT(15X,'EMISSION RATE =',2X,F10.5,1X,'GR/DSCF'/)
      WRITE(3,220)XISO
220   FORMAT(15X,'ISOKINETIC SAMPLING VARIATION =',2X,F5.
    1 0,1X,'%'/////)
      RETURN
      END
```

APPENDIX E
SOURCE SAMPLING CALCULATIONS

SUMMARY OF EMISSION TEST EQUATIONS

Method 1. Site Selection

Equivalent diameter for rectangular duct:

$$D_E = \frac{2 \times L \times W}{L + W}$$

where L and W are the duct dimensions.

Method 2. Gas Velocity and Flow Rate

S-type pitot coefficient:

$$C_{p(s)} = C_{p(std)} \times \sqrt{\frac{\Delta P_{std}}{\Delta P_s}}$$

Average deviation of pitot coefficient from the mean for that side:

$$\delta = \sum \left(\frac{\left| C_{p(s)} - C_{p(A\ or\ B)} \right|}{3} \right)$$

Gas velocity:

$$V_s = K_p \times C_p \times \left(\frac{T_{gas}}{P_{stat} \times M_{gas}}\right)^{\frac{1}{2}} \times \left(\Delta P_p\right)^{\frac{1}{2}} \quad (5.7b)$$

Exhaust gas volume flow rate:

$$Q_s = 3600 \times \left(1 - B_{H_2O}\right) \times v_s \times A_s \times \left(\frac{P_{stat}}{P_{std}} \times \frac{T_{std}}{T_{gas}}\right)$$

where the units of Q_s are dscmh or dscfh.

Method 3. Molecular Weight of Exhaust Gas

Dry molecular weight of mixture:

$$M_{dry} = \sum B_i \times M_i$$

Total molecular weight:

$$M_{gas} = M_{dry} \times \left(1 - B_{H_2O}\right) + 18B_{H_2O}$$

Percent excess air:

$$\%EA = \frac{\left(\%O_2\right) - 0.5\,(\%CO)}{0.264\left(\%N_2\right) - \left(\%O_2\right) + 0.5\,(\%CO)} \times 100$$

Method 4. Moisture Content of Exhaust Gas

Volume, at standard conditions, of water vapor condensed:

$$V_{cond} = \frac{V_{liq} \times \rho \times R_u \times T_{std}}{P_{std} \times 18}$$

Volume, at standard conditions, of water vapor adsorbed:

$$V_{sg} = \frac{m_{sg} \times R_u \times T_{std}}{P_{std} \times 18}$$

Water vapor fraction:

$$B_{H_2O} = \frac{V_{cond} + V_{sg}}{V_{cond} + V_{sg} + V_{m(std)}}$$

Source Sampling

Dry Gas Meter calibration:

$$Y = \frac{V_{wtm} + C_{wtm} \times T_{dgm} \times P_{bar}}{V_{wtm} + T_{wtm} \times \left(P_{bar} + \frac{\Delta H}{13.6}\right)}$$

Orifice Calibration:

$$\Delta H@ = 0.00117 \times \frac{\Delta H}{P_{bar} \times T_i} \times \left(\frac{T_{wtm} \times \theta}{V_{wtm} \times C_{wtm}}\right)^2 \text{ (SI)} \quad \textbf{(11.2a)}$$

$$\Delta H@ = 0.0317 \times \frac{\Delta H}{P_{bar} \times T_i} \times \left(\frac{T_{wtm} \times \theta}{V_{wtm} \times C_{wtm}}\right)^2 \text{ (US)} \quad \textbf{(11.2b)}$$

Volume, at dry standard conditions, of gas sampled:

$$V_{m(std)} = V_m \times Y \times \frac{P_m}{P_{std}} \times \frac{T_{std}}{T_m} = V_m \times Y \times \frac{\left(P_{bar} + \frac{\Delta H_{avg}}{13.6}\right)}{T_{m(avg)}} K_1 \quad \textbf{(13.1)}$$

Isokinetic Calculations

Orifice head loss for isokinetic sampling rate:

$$\Delta H_{iso} = \left[K_p^2 \times \left(\frac{\pi}{4}\right)^2 \times C_p^2\right]$$

$$\times \left[D_n^4 \times \left(\frac{\Delta H@}{0.0331}\right) \times \left(1 - B_{H_2O}\right)^2 \times \frac{M_d}{M_{gas}} \times \frac{P_{stat}}{P_m} \times \frac{T_m}{T_{gas}}\right] \times \Delta P_p \text{ (SI)} \quad \textbf{(12.8)}$$

Nozzle diameter:

$$\Delta H_{iso} = \left[8.204 \times 10^{-5} \times D_n^4 \times \Delta H@ \times C_p^2 \times \left(1 - B_{H_2O}\right)^2 \times \frac{M_d}{M_{gas}} \times \frac{T_m}{T_{gas}} \times \frac{P_{stat}}{P_m}\right] \times \Delta P_p \quad \text{(SI)} \tag{12.9}$$

and for U.S. Customary units, with nozzle diameter in inches:

$$\Delta H_{iso} = \left[846.7 D_n^4 \times \Delta H@ \times C_p^2 \times \left(1 - B_{H_2O}\right)^2 \times \frac{M_d}{M_{gas}} \times \frac{T_m}{T_{gas}} \times \frac{P_{stat}}{P_m}\right] \times \Delta P_p \quad \text{(US)} \tag{12.9a}$$

Percent of isokinetic:

$$\%I = \frac{100 \times V_m \times YP_m}{\left(1 - B_{H_2O}\right) \times T_m \times A_n \times \theta \times K_p \times C_p} \times \left(\frac{T_{gas} \times M_{gas}}{P_{stat}}\right)^{\frac{1}{2}} \times \left(\Delta P_p\right)^{\frac{1}{2}}_{avg} \tag{12.16}$$

or

$$\%I = \frac{100 \times V_m \times YP_m \times T_{gas}}{v_s \times A_n \times \theta \times P_{stat} \times T_m \times \left(1 - B_{H_2O}\right) \times 60} \tag{12.17}$$

where the factor of 60 is included because v_s is generally in feet or meters per second and Q in minutes.

OR, from "intermediate data":

$$\%I = \frac{100 \times T_{gas} \times V_{m(std)} \times P_{std}}{60 \times T_{std} \times v_s \times \theta \times A_n \times P_{stat} \times \left(1 - B_{H_2O}\right)} \tag{12.18}$$

Emission Rate Calculations

Concentration, at dry standard conditions:

$$C_s = \frac{m_s}{V_{m(std)}} \tag{13.2}$$

Actual concentration:

$$C_{s(act)} = C_s \times \left(\frac{P_{stat}}{P_{std}} \times \frac{T_{std}}{T_{gas}} \right) \times \left(1 - B_{H_2O}\right) \qquad \textbf{(13.3)}$$

Concentration adjusted to 50% excess air:

$$C_{s(50\%EA)} = C_s \times \left(\frac{100 + \%EA}{150} \right) \qquad \textbf{(13.4)}$$

$$C_{s(50\%EA)} = C_s \times \left(\frac{21}{21 - 1.5\left[\%O_2\right] - 0.75[\%CO] - 0.133\left[\%N_2\right]} \right)$$

(13.5)

$$C_{s(50\%EA)} = C_s \times \frac{B_{CO_2} + 1.75}{2.88 - 13.9 * B_{O_2}} \qquad \textbf{(13.6)}$$

Concentration adjusted to 12% CO_2:

$$C_{s(12\%CO_2)} = C_s \times \left(\frac{12}{\%CO_2} \right) \qquad \textbf{(13.7)}$$

Mass emission rate:

$$\frac{dm_p}{dt} = \dot{m}_p = C_s \times Q_{s(dry,std)} \qquad \textbf{(13.8)}$$

Emission rate per process parameter:

$$ER = \frac{\dot{m}_p}{PR} = \frac{C_s \times Q_{s(dry,std)}}{PR} \qquad \textbf{(13.9)}$$

Emission rate per unit energy consumption:

$$ER = \frac{C_s \times Q_{s(dry,std)}}{Q_H} \qquad \textbf{(13.10)}$$

F-factor emission rate:

$$ER = C_s \times F_d \times \left(\frac{20.9}{20.9 - \%O_2} \right) \qquad \textbf{(13.20)}$$

APPENDIX F
HIGH VOLUME TECHNIQUE

HIGH VOLUME SAMPLER CALIBRATION METHOD 40CFR50, APPENDIX B

9.0 *Calibration*

9.1 Calibration of the high volume sampler's flow indicating or control device is necessary to establish traceability of the field measurement to a primary standard via a flow rate transfer standard. Determination of the corrected flow rate from the sampler flow indicator is addressed in Section 10.1.

Note: The following calibration procedure applies to a conventional orifice-type flow transfer standard and an orifice-type flow indicator in the sampler (the most common types). For samplers using a pressure recorder having a square-root scale, 3 other acceptable calibration procedures are provided in Reference 12. Other types of transfer standards may be used if the manufacturer or user provides an appropriately modified calibration procedure that has been approved by EPA under Section 2.8 of Appendix C to Part 58 of this chapter.

9.2 *Certification of the flow rate transfer standard.*

9.2.1 *Equipment required:* Positive displacement standard volume meter traceable to the National Bureau of Standards (such as a Roots meter or equivalent), stop-watch, manometer, thermometer, and barometer.

9.2.2 Connect the flow rate transfer standard to the inlet of the standard volume meter. Connect the manometer to measure the pressure at the inlet of the standard volume meter. Connect the orifice manometer to the pressure tap on the transfer standard. Connect a high-volume air pump (such as a high-volume sampler blower) to the outlet side of the standard volume meter.

9.2.3 Check for leaks by temporarily clamping both manometer lines (to avoid fluid loss) and blocking the orifice with a large-diameter rubber stopper, wide cellophane tape, or other suitable means. Start the high-volume air pump and note any change in the standard volume meter reading. The reading should remain constant. If the reading changes, locate any leaks by listening for a whistling sound and/or retightening all connections, making sure that all gaskets are properly installed.

9.2.4 After satisfactorily completing the leak check as described above, unclamp both manometer lines and zero both manometers.

9.2.5 Achieve the appropriate flow rate through the system, either by means of the variable flow resistance in the transfer standard or by varying the voltage to the air pump. (Use of resistance plates as shown in Figure 1a is discouraged because the above leak check must be repeated each time a new resistance plate is installed.) At least five different but constant flow rates, evenly distributed, with at least three in the specified flow rate interval (1.1 to 1.7 m^3/min [39–60 ft^3/min]), are required.

9.2.6 Measure and record the certification data on a form similar to the one illustrated in Figure 1 according to the following steps.

9.2.7 Observe the barometric pressure and record as P_1 (item 8 in Figure 1).

9.2.8 Read the ambient temperature in the vicinity of the standard volume meter and record it as T_1 (item 9 in Figure 1).

9.2.9 Start the blower motor, adjust the flow, and allow the system to run for at least 1 min for a constant motor speed to be attained.

9.2.10 Observe the standard volume meter reading and simultaneously start a stopwatch. Record the initial meter reading (V_i) in column 1 of Figure 1.

9.2.11 Maintain this constant flow rate until at least 3 m^3 of air have passed through the standard volume meter. Record the standard volume meter inlet pressure manometer reading as ΔP (column 5 in Figure 1), and the orifice manometer reading as ΔH (column 7 in Figure 1). Be sure to indicate the correct units of measurement.

9.2.12 After at least 3 m^3 of air have passed through the system, observe the standard volume meter reading while simultaneously stopping the stopwatch. Record the final meter reading (V_f) in column 2 and the elapsed time (t) in column 3 of Figure 1.

9.2.13 Calculate the volume measured by the standard volume meter at meter conditions of temperature and pressures as $V_m = V_f - V_i$. Record in column 4 of Figure 1.

9.2.14 Correct this volume to standard volume (std m3) as follows:

$$V_{std} = V_m \times \frac{P_1 - \Delta P}{P_{std}} \times \frac{T_{std}}{T_1}$$

Where:

V_{std} = standard volume, std m3

V_m = actual volume measured by the standard volume meter;

P_1 = barometric pressure during calibration, mm Hg or kPa;

ΔP = differential pressure at inlet to volume meter, mm Hg or kPa;

P_{std} = 760 mm Hg or 101 kPa;

T_{std} = 298 K;

T_1 = ambient temperature during calibration, K.

Calculate the standard flow rate (std m^3/min) as follows:

$$Q_{std} = \frac{V_{std}}{t}$$

where:

Q_{std} = standard volumetric flow rate, std m^3/min

t = elapsed time, minutes

Record Q_{std} to the nearest 0.01 std m^3/min in column 6 of Figure 1.

9.2.15 Repeat steps 9.2.9 through 9.2.14 for at least four additional constant flow rates, evenly spaced over the approximate range of 1 to 1.8 std m^3/min (35–64 ft^3/min).

9.2.16 For each flow, compute

$$\sqrt{\Delta\Delta H} \times (P_1/P_{std}) \times (298/T_1)$$

(column 7a of Figure 1) and plot these values against Q_{std}. Be sure to use consistent units (mm Hg or kPa) for barometric pressure. Draw the orifice transfer standard certification curve or calculate the linear least squares slope (m) and intercept (b) or the certification curve:

$$\sqrt{\Delta\Delta H} \times (P_1/P_{std}) \times (298/T_1) = m \times Q_{std} + b$$

See Figure 1. A certification graph should be readable to 0.02 std m^3/min.

9.2.17 Recalibrate the transfer standard annually or as required by applicable quality control procedures. (See Reference 2.)

9.3 *Calibration of sampler flow indicator.*

Note: For samplers equipped with a flow controlling device, the flow controller must be disabled to allow flow changes during calibration of the sampler's flow indicator, or the alternate calibration of the flow controller given in 9.4 may be used. For samplers using an orifice-type flow indicator downstream of the motor, do not vary the flow rate by adjusting the voltage or power supplied to the sampler.

9.3.1 A form similar to the one illustrated in Figure 2 should be used to record the calibration data.

9.3.2 Connect the transfer standard to the inlet of the sampler. Connect the orifice manometer to the orifice pressure tap. Make sure there are no leaks between the orifice unit and the sampler.

9.3.3 Operate the sampler for at least 5 minutes to establish thermal equilibrium prior to the calibration.

9.3.4 Measure and record the ambient temperature, T_2, and the barometric pressure, P_2, during calibration.

9.3.5 Adjust the variable resistance or, if applicable, insert the appropriate resistance plate (or no plate) to achieve the desired flow rate.

9.3.6 Let the sampler run for at least 2 min to re-establish the run-temperature conditions. Read and record the pressure drop across the orifice (ΔH) and the sampler flow rate indication (I) in the appropriate columns of Figure 2.

9.3.7 Calculate $\sqrt{\Delta H \times (P_2/P_{std}) \times (298/T_2)}$ and determine the flow rate at standard conditions (Q_{std}) either graphically from the certification curve or by calculating Q_{std} from the least square slope and the intercept of the transfer standard's transposed certification curve: $Q_{std} = 1/m \times \left[\sqrt{\Delta H \times (P_2/P_{std} \times (298/T_2)} - b\right]$. Record the value of Q_{std} on Figure 2.

9.3.8 Repeat steps 9.3.5, 9.3.6, and 9.3.7 for several additional flow rates distributed over a range that includes 1.1 to 1.7 std m^3/min.

9.3.9 Determine the calibration curve by plotting values of the appropriate expression involving I, selected from Table 1, against Q_{std}. The choice of expression from Table 1 depends on the flow rate measurement device used (see Section 7.4.1) and also on whether the calibration curve is to incorporate geographic average barometric pressure (P_a) and seasonal average temperature (T_a) for the site to approximate actual pressure and temperature. Where P_a and T_a can be determined for a site for a seasonal period such that the actual barometric pressure and temperature at the site do not vary by more than ± 60 mm Hg (8 kPa) from Pa or ± 15 °C from T_a, respectively, then using P_a and T_a avoids the need for subsequent pressure and temperature calculation when the sampler is used. The geographic average barometric pressure (P_a) may be estimated from an altitude-pressure table or by making an (approximate) elevation correction of –26 mm Hg (–3.46 kPa) for each 305 m (1000 ft) above sea level (760 mm Hg or 101 kPa). The seasonal average temperature (T_a) may be estimated from weather station or other records. Be sure to use consistent units (mm Hg or kPa) for barometric pressure.

9.3.10 Draw the sampler calibration curve or calculate the linear least squares slope (m), intercept (b), and correlation coefficient of the calibration curve: [Expression from Table 1] = $mQ_{std} + b$. See Figure 2. Calibration curves should be readable to 0.02 std m^3/min.

9.3.11 For a sampler equipped with a flow controller, the flow controlling mechanism should be re-enabled and set to a flow near the lower flow limit to allow maximum control range. The sample flow rate should be verified at this

Table 1. Expressions for Plotting Sampler Calibration Curves

Type of sampler flow rate measuring device	Expression: For actual pressure and temperature corrections	Expression: For incorporation of geographic average pressure and seasonal average temperature
Mass flowmeter	I	I
Orifice and pressure indicator	$\sqrt{I \times \left(\frac{P_2}{P_{std}}\right) \times \left(\frac{298}{T_2}\right)}$	$\sqrt{I \times \left(\frac{P_2}{P_a}\right) \times \left(\frac{T_a}{T_2}\right)}$
Rotameter, or orifice and pressure recorder having square root scale[a]	$I \times \sqrt{\left(\frac{P_2}{P_{std}}\right) \times \left(\frac{298}{T_2}\right)}$	$I \times \sqrt{\left(\frac{P_2}{P_a}\right) \times \left(\frac{T_a}{T_2}\right)}$

[a] This scale is recognizable by its nonuniform divisions and is the most commonly available for high-volume samplers.

Table 2. Expressions for Determining Flow Rate During Sampler Operations

Type of sampler flow rate measuring device	Expression: For actual pressure and temperature corrections	Expression: For use when geographic average pressure and seasonal average temperature have been incorporated into the sampler calibration
Mass flowmeter	I	I
Orifice and pressure indicator	$\sqrt{I \times \left(\frac{P_2}{P_{std}}\right) \times \left(\frac{298}{T_2}\right)}$	$\sqrt{I}$
Rotameter, or orifice and pressure recorder having square root scale[a]	$I \times \sqrt{\left(\frac{P_2}{P_{std}}\right) \times \left(\frac{298}{T_2}\right)}$	I

[a] This scale is recognizable by its nonuniform divisions and is the most commonly available for high-volume samplers.

time with a clean filter installed. Then add two or more filters to the sampler to see if the flow controller maintains a constant flow; this is particularly important at high altitudes where the range of the flow controller may be reduces.

9.4 *Alternate calibration of flow-controlled samplers.*

A flow-controlled sampler may be calibrated solely at its controlled flow rate, provided that previous operating history of the sampler demonstrates that the flow rate is stable and reliable. In this case, the flow indicator may remain uncalibrated but should be used to indicate any relative change between initial and final flows, and the sampler should be recalibrated more often to minimize potential loss of samples because of controller malfunction.

9.4.1 Set the flow controller for a flow near the lower limit of the flow range to allow maximum control range.

9.4.2 Install a clean filter in the sampler and carry out steps 9.9.2, 9.3.3, 9.3.4, 9.3.6, and 9.3.7.

9.4.3 Following calibration, add one or two additional clean filters to the sampler, reconnect the transfer standard, and operate the sampler to verify that the controller maintains the same calibrated flow rate; this is particularly important at high altitudes where the flow control range may be reduced.

ORIDICE TRANSFER STANDARD CERTIFICATION WORKSHEET

	(1)	(2)	(3)	(4)	(5)	(6)	(7)	(7a)
Run No.	Meter reading tart V_1 (m^3)	Meter reading stop V_f (m^3)	Sampling time t (min)	Volume measured V_m (m^3)	Differential pressure (at inlet to volume meter) ΔP (mm Hg or kPa)	(X) flow rate Q_{std} (std m^3/min)	Pressure drop across orifice ΔH ☐ (in) or ☐ of water	(Y) $\sqrt{\Delta H \times \left(\frac{P_1}{P_{std}}\right) \times \left(\frac{298}{T_1}\right)}$
1								
2								
3								
4								
5								
6								

RECORDED CALIBRATION DATA

Standard volume meter no. ______

Transfer standard type: ☐ orifice ☐ other

Model No. ______ Serial No. ______

(8) P_1: ______ mm Hg (or kPa) (10) P_{std}: 760 mm Hg (or 101 kPa)

(9) T_1: ______ K (11) T_{std}: 298 K

Calibration performed by: ______

Date: ______

CALCULATION EQUATIONS

(1) $V_m \times V_f - V_i$

(2) $V_{std} = V_m \times \left(\frac{P_1 - \Delta P}{P_{std}}\right) \times \left(\frac{T_{std}}{T_1}\right)$

(3) $Q_{std} = \frac{V_{std}}{t}$

LEAST SQUARES CALCULATIONS

Linear (Y + mX + b) regression equation of $Y = \sqrt{\Delta H \times (P_1/P_{std}) \times (298/T_1)}$ on $X = Q_{std}$ for Orifice Calbration Unit (i.e., $\sqrt{\Delta H \times (P_1/P_{std}) \times (298/T_1)} = (m \times Q_{std} + b)$

Slope (m) = ______ Intercept (b) × ______ Correlation coefficient (r) = ______

To use for subsequent calibration: $X = \frac{1}{m} \times (Y - b)$: $Q_{std} = \frac{1}{m} \times \left(\sqrt{\left\{\Delta H \times \left(\frac{P_2}{P_{std}}\right) \times \left(\frac{298}{T_1}\right)\right\}} - b\right)$

Figure 1. Example of orifice transfer standard certification worksheet.

HIGH-VOLUME AIR SAMPLER CALIBRATION WORKSHEET

Site Locations: ______________________________

Date: ______________ Barometric Pressure P_2 mm Hg (or kPa) ______________

Calibrated By: ______________ Temperature, T_2 (K) ______________

Sampler No. ______________ Serial No. ______________

Transfer std. type: ______________ Serial No. ______________

P_{std} = 760 mm HG (or 101)

Optional:

Average barometric pressure: P_a ______________

Seasonal average temperature: T_a ______________

No.	ΔH Pressure drop across orifice ☐ (in) or ☐ (cm) of water	$\sqrt{\Delta H \times \left(\frac{P_2}{P_{std}}\right) \times \left(\frac{298}{T_2}\right)}$	(X) Q_{std} (from orifice certification) std m³/min	I Sampler flow rate indication (arbitrary)	For specific pressure and temperature corrections (see Table 1): ☐ I or ☐ $\sqrt{I \times \left(\frac{P_2}{P_{std}}\right) \times \left(\frac{298}{T_2}\right)}$ or ☐ $I \times \sqrt{\left(\frac{P_2}{P_{std}}\right) \times \left(\frac{298}{T_2}\right)}$	(Y) For Incorporation of average pressure and seasonal average temperature (see Table I): ☐ I or ☐ $\sqrt{I \times \left(\frac{P_2}{P_a}\right) \times \left(\frac{T_a}{T_2}\right)}$ or ☐ $I \times \sqrt{\left(\frac{P_2}{P_a}\right) \times \left(\frac{T_a}{T_2}\right)}$
1						
2						
3						
4						
5						
6						

LEAST SQUARES CALCULATIONS

Linear regression of Y on X: Y = mX + b; Y = appropriate expression from Table 1; X = Q_{std}

Slope (m) = __________ Intercept (b) = __________ Correlation Coeff. (r) = __________

To determine subsequent flow rate during use: $X = \frac{1}{m} \times (Y - b)$; $Q_{std} = \frac{1}{m}$ ([appropriate expression from Table 2] − b)

Figure 2. Example of high-volume air sampler calibration worksheet.

APPENDIX G1 SATURATION PRESSURE TABLE

SATURATION VAPOR PRESSURE OF WATER, mm HG

Temp °C	0.0	0.2	0.4	0.6	0.8
−15	1.436	1.414	1.390	1.368	1.345
−14	1.56	1.534	1.511	1.485	1.460
−13	1.691	1.665	1.637	1.611	1.585
−12	1.834	1.804	1.776	1.748	1.720
−11	1.987	1.955	1.924	1.893	1.863
−10	2.149	2.116	2.084	2.050	2.018
−9	2.326	2.289	2.254	2.219	2.184
−8	2.514	2.475	2.437	2.399	2.362
−7	2.715	2.674	2.663	2.593	2.553
−6	2.931	2.887	2.843	2.800	2.757
−5	3.163	3.115	3.069	3.022	2.976
−4	3.41	3.359	3.309	3.259	3.211
−3	3.675	3.620	3.567	3.514	3.461
−2	3.956	3.898	3.841	3.785	3.730
−1	4.258	4.196	4.135	4.075	4.016
0	4.579	4.513	4.448	4.385	4.320
0	4.579	4.647	4.715	4.785	4.855
1	4.926	4.998	5.070	5.144	5.219
2	5.294	5.370	5.447	5.525	5.605
3	5.685	5.766	5.848	5.931	6.015
4	6.101	6.187	6.274	6.363	6.453
5	6.543	6.635	6.728	6.822	6.917

6	7.013	7.111	7.209	7.309	7.411
7	7.513	7.617	7.772	7.828	7.936
8	8.045	8.155	8.267	8.380	8.494
9	8.609	8.727	8.845	8.965	9.086
10	9.209	9.333	9.458	9.585	9.714
11	9.844	9.976	10.109	10.244	10.380
12	10.518	10.658	10.799	10.941	11.085
13	11.231	11.379	11.528	11.680	11.833
14	11.987	12.144	12.302	12.462	12.624
15	12.788	12.953	13.121	13.290	13.461
16	13.634	13.809	13.987	14.166	14.347
17	14.53	14.715	14.903	15.092	15.284
18	15.477	15.673	15.871	16.071	16.272
19	16.477	16.685	16.894	17.105	17.319
20	17.535	17.753	17.974	18.197	18.422
21	18.65	18.880	19.113	19.349	19.587
22	19.827	20.070	20.316	20.565	20.815
23	21.068	21.324	21.583	21.845	22.110
24	22.377	22.648	22.922	23.198	23.476
25	23.756	24.039	24.326	24.617	24.912
26	25.209	25.509	25.812	26.117	26.426
27	26.739	27.055	27.374	27.696	28.021
28	28.349	28.680	29.015	29.354	29.697
29	30.043	30.392	30.754	31.120	31.461
30	31.824	32.191	32.561	32.934	33.312
31	33.695	34.082	34.471	34.864	35.261
32	35.663	36.068	36.477	36.891	37.308
33	37.729	38.155	38.584	39.018	39.457
34	39.898	40.344	40.796	41.251	41.710
35	42.175	42.644	43.117	43.595	44.078
36	44.563	45.054	45.549	46.050	46.556
37	47.067	47.582	48.102	48.627	49.157
38	49.692	50.231	50.774	51.323	51.879
39	52.442	53.009	53.580	54.156	54.737
40	55.324	55.91	56.51	57.11	57.72
41	58.34	58.96	59.58	60.22	60.86
42	61.50	62.14	62.80	63.46	64.12
43	64.80	65.48	66.16	66.86	67.56
44	68.26	68.97	69.69	70.41	71.14
45	71.88	72.62	73.36	74.12	74.88
46	75.65	76.43	77.21	78.00	78.80
47	79.60	80.41	81.23	82.05	82.87
48	83.71	84.56	85.42	86.28	87.14
49	88.02	88.90	89.89	90.69	91.59
50	92.51	93.4	94.4	95.3	96.3
51	97.20	98.2	99.2	100.1	101.1
52	102.09	103.1	104.1	105.2	106.2
53	107.20	108.3	109.3	110.4	111.4

54	112.51	113.6	114.7	115.8	116.9
55	118.04	119.2	120.3	121.5	122.6
56	123.80	125.0	126.2	127.4	128.6
57	129.82	131.1	132.3	133.6	134.7
58	136.08	137.4	138.7	140.0	141.2
59	142.60	144.0	145.3	146.7	148.0
60	149.38	150.7	152.1	153.5	155.0
61	156.43	157.8	159.3	160.8	162.3
62	163.77	165.2	166.7	168.3	169.9
63	171.38	172.9	174.5	176.1	177.7
64	179.31	180.9	182.5	184.2	185.9
65	187.54	189.2	190.9	192.6	194.4
66	196.09	197.8	199.5	201.3	203.2
67	204.96	206.7	208.6	210.4	212.3
68	214.17	216.1	217.9	219.8	221.8
69	223.73	225.7	227.6	229.6	231.7
70	233.7	235.7	237.7	239.7	241.9
71	243.9	246.0	248.1	250.2	252.5
72	254.6	256.8	258.9	261.1	263.5
73	265.7	268.0	270.2	272.5	274.9
74	277.2	279.6	281.8	284.2	286.7
75	289.1	291.6	293.9	296.4	298.9
76	301.4	303.9	306.4	308.9	311.6
77	314.1	316.7	319.2	321.9	324.7
78	327.3	330.0	332.6	335.4	338.3
79	341.0	343.8	346.5	349.3	352.3
80	355.1	358.0	360.8	363.7	366.8
81	369.7	372.7	375.6	378.7	381.9
82	384.9	388.0	391.0	394.2	397.5
83	400.6	403.8	506.9	410.2	413.6
84	416.8	420.2	423.4	426.7	430.2
85	433.6	437.1	440.3	443.8	447.4
86	450.9	454.5	457.8	461.4	465.1
87	468.7	472.4	475.9	479.6	483.4
88	487.1	490.9	494.5	498.3	502.3
89	506.1	510.0	513.8	517.7	521.8
90	525.76	529.8	533.7	537.7	542.0
91	546.05	550.2	554.2	558.4	562.8
92	566.99	571.3	575.4	579.7	584.3
93	588.60	593.1	597.3	601.8	606.4
94	610.90	615.5	619.9	624.5	629.3
95	633.90	638.6	643.2	647.9	652.9
96	657.62	662.5	667.2	672.0	677.2
97	682.07	687.1	691.9	696.9	702.2
98	707.27	712.5	717.4	722.6	728.0
99	733.24	738.6	743.7	749.0	754.6
100	760.00	765.5	770.8	776.3	782.1
101	787.57	793.2	798.8	804.5	810.2

APPENDIX G2
PSYCHROMETRY

Barometric Pressure 29.92 In. Hg

ADDITIVE CORRECTIONS FOR W, h, AND v WHEN BAROMETRIC PRESSURE DIFFERS FROM STANDARD BAROMETER

Wet Bulb Temp. t′	Sat. Vapor Press. In. Hg	−900		900		1800		2700		3700		4800		5800	
		$\Delta p = +1$		$\Delta p = -1$		$\Delta p = -2$		$\Delta p = -3$		$\Delta p = -4$		$\Delta p = -5$		$\Delta p = -6$	
		$\Delta W'_s$	Δh	$\Delta W'_s$	Δh	$\Delta W'_s$	Δh	$\Delta W'_s$	Δh	$\Delta W'_s$	Δh	$\Delta W'_s$	Δh	$\Delta W'_s$	Δh
60	0.522	−2.5	−0.40	2.7	0.42	5.7	0.88	8.8	1.37	12.2	1.90	15.9	2.47	19.9	3.09
61	0.540	−2.6	−0.41	2.8	0.44	5.9	0.91	9.2	1.43	12.7	1.98	16.5	2.57	20.7	3.22
62	0.560	−2.7	−0.43	2.9	0.46	6.1	0.95	9.5	1.48	13.2	2.06	17.1	2.66	21.4	3.33
63	0.580	−2.8	−0.44	3.0	0.48	6.3	0.98	9.9	1.54	13.7	2.13	17.7	2.76	22.3	3.47
64	0.601	−2.9	−0.46	3.2	0.49	6.5	1.02	10.3	1.59	14.2	2.21	18.4	2.87	23.1	3.60
65	0.622	−3.1	−0.48	3.3	0.51	6.8	1.06	10.6	1.65	14.7	2.29	19.1	2.98	23.9	3.73
66	0.644	−3.2	−0.50	3.4	0.53	7.1	1.10	11.0	1.72	15.3	2.38	19.8	3.09	24.8	3.87
67	0.667	−3.3	−0.51	3.5	0.55	7.3	1.14	11.4	1.78	15.8	2.47	20.5	3.20	25.7	4.01
68	0.690	−3.4	−0.53	3.7	0.57	7.6	1.18	11.8	1.84	16.4	2.56	21.3	3.32	26.7	4.16
69	0.714	−3.5	−0.55	3.8	0.59	7.9	1.23	12.2	1.90	17.0	2.65	22.1	3.45	27.7	4.32
70	0.739	−3.7	−0.57	3.9	0.61	8.1	1.27	12.7	1.98	17.6	2.75	22.9	3.58	28.7	4.48
71	0.765	−3.8	−0.59	4.1	0.64	8.4	1.32	13.1	2.05	18.2	2.84	23.7	3.70	29.7	4.64
72	0.791	−3.9	−0.61	4.2	0.66	8.7	1.36	13.6	2.13	18.8	2.94	24.6	3.84	30.9	4.82
73	0.818	−4.1	−0.63	4.4	0.69	9.0	1.41	14.1	2.20	19.5	3.05	25.5	3.99	31.9	4.99
74	0.846	−4.2	−0.66	4.6	0.71	9.4	1.46	14.6	2.28	20.2	3.16	26.4	4.14	33.1	5.18
75	0.875	−4.4	−0.68	4.7	0.74	9.7	1.52	15.1	2.36	20.9	3.27	27.4	4.28	34.3	5.37
76	0.905	−4.5	−0.71	4.9	0.77	10.0	1.57	15.7	2.46	21.7	3.39	28.3	4.42	35.5	5.56
77	0.935	−4.7	−0.73	5.1	0.79	10.4	1.63	16.3	2.55	22.5	3.52	29.4	4.61	36.9	5.77
78	0.967	−4.9	−0.76	5.2	0.82	10.8	1.69	16.9	2.65	23.3	3.65	30.5	4.77	38.2	5.98
79	0.999	−5.0	−0.79	5.4	0.85	11.2	1.75	17.5	2.74	24.2	3.79	31.6	4.95	39.6	6.20
80	1.032	−5.2	−0.82	5.6	0.88	11.6	1.82	18.1	2.84	25.1	3.93	32.7	5.13	41.0	6.43
81	1.067	−5.4	−0.85	5.8	0.91	12.0	1.88	18.8	2.95	26.0	4.08	33.9	5.32	42.5	6.66
82	1.102	−5.6	−0.88	6.0	0.94	12.5	1.96	19.5	3.06	27.0	4.24	35.1	5.51	44.0	6.90
83	1.138	−5.8	−0.91	6.2	0.97	12.9	2.02	20.2	3.17	28.0	4.39	36.4	5.71	45.6	7.15
84	1.175	−6.0	−0.94	6.4	1.00	13.3	2.10	20.9	3.28	28.9	4.54	37.7	5.92	47.2	7.41
85	1.214	−6.2	−0.97	6.7	1.05	13.8	2.17	21.6	3.39	29.9	4.69	39.0	6.12	48.9	7.67
86	1.253	−6.4	−1.00	6.9	1.08	14.3	2.24	22.3	3.50	30.9	4.85	40.4	6.34	50.6	7.94
87	1.294	−6.7	−1.05	7.1	1.11	14.8	2.32	23.1	3.63	32.0	5.02	41.8	6.56	52.3	8.21
88	1.335	−6.9	−1.08	7.4	1.16	15.3	2.40	23.9	3.75	33.1	5.20	43.2	6.79	54.2	8.51
89	1.378	−7.1	−1.12	7.7	1.21	15.9	2.50	24.8	3.90	34.3	5.39	44.8	7.04	56.2	8.83
90	1.422	−7.4	−1.16	7.9	1.24	16.5	2.59	25.7	4.04	35.6	5.60	46.4	7.29	58.2	9.15
91	1.467	−7.6	−1.20	8.2	1.29	17.0	2.67	26.6	4.18	36.9	5.80	48.1	7.56	60.3	9.48
92	1.514	−7.9	−1.24	8.5	1.34	17.6	2.77	27.5	4.33	38.2	6.01	49.8	7.83	62.5	9.83
93	1.561	−8.2	−1.29	8.8	1.39	18.3	2.88	28.5	4.49	39.6	6.23	51.5	8.11	64.7	10.18
94	1.610	−8.5	−1.34	9.1	1.43	18.9	2.98	29.5	4.64	41.0	6.46	53.4	8.41	67.0	10.55
95	1.661	−8.8	−1.39	9.4	1.48	19.6	3.09	30.5	4.80	42.4	6.68	55.2	8.69	69.3	10.92
96	1.712	−9.1	−1.43	9.8	1.54	20.2	3.18	31.5	4.96	43.8	6.90	57.2	9.01	71.7	11.30
97	1.766	−9.4	−1.48	10.1	1.59	20.9	3.29	32.6	5.14	45.3	7.14	59.2	9.33	74.2	11.70
98	1.820	−9.7	−1.53	10.4	1.64	21.7	3.42	33.8	5.33	47.0	7.41	61.3	9.67	76.8	12.11
99	1.876	−10.1	−1.59	10.8	1.70	22.4	3.53	35.0	5.52	48.6	7.67	63.5	10.02	79.6	12.56
100	1.933	−10.4	−1.64	11.2	1.77	23.2	3.66	36.3	5.73	50.4	7.96	65.7	10.37	82.5	13.02
101	1.992	−10.8	−1.71	11.6	1.83	24.0	3.79	37.6	5.94	52.2	8.24	68.0	10.74	85.4	13.48
102	2.053	−11.1	−1.75	12.0	1.90	24.8	3.92	38.9	6.14	54.1	8.54	70.5	11.13	88.5	13.98
103	2.115	−11.5	−1.82	12.4	1.96	25.7	4.06	40.3	6.37	56.0	8.85	72.9	11.52	91.6	14.47
104	2.179	−11.9	−1.88	12.8	2.02	26.6	4.20	41.6	6.58	57.9	9.15	75.5	11.93	94.8	14.98

Psychrometric Chart. Reprinted with permission of the Carrier Corporation, Syracuse, New York.

105	2.244	−12.4	−1.96	13.3	2.10	27.6	4.36	43.1	6.82	59.9	9.47	78.2	12.37	98.2	15.53
106	2.311	−12.8	−2.02	13.7	2.17	28.6	4.52	44.6	7.06	62.1	9.82	81.1	12.83	101.7	16.09
107	2.381	−13.2	−2.09	14.2	2.25	29.6	4.68	46.2	7.31	64.3	10.17	84.1	13.31	105.3	16.66
108	2.450	−13.7	−2.17	14.7	2.33	30.6	4.84	47.7	7.55	66.5	10.53	87.0	13.77	109.1	17.27
109	2.523	−14.2	−2.25	15.3	2.42	31.7	5.02	49.4	7.82	68.9	10.91	90.0	14.25	113.0	17.90
110	2.597	−14.7	−2.33	15.8	2.50	32.8	5.20	51.3	8.12	71.3	11.30	93.1	14.75	117.0	18.54
111	2.673	−15.2	−2.41	16.3	2.58	34.0	5.39	53.1	8.42	73.8	11.70	96.4	15.28	121.4	19.24
112	2.751	−15.7	−2.49	16.9	2.68	35.2	5.58	55.0	8.72	76.4	12.11	99.9	15.84	125.9	19.96
113	2.831	−16.3	−2.58	17.5	2.78	36.4	5.77	56.9	9.03	79.2	12.56	103.5	16.42	130.4	20.68
114	2.913	−16.9	−2.68	18.1	2.87	37.7	5.98	58.9	9.35	82.0	13.01	107.3	17.03	135.0	21.42
115	2.996	−17.4	−2.76	18.8	2.96	39.1	6.21	61.0	9.68	85.0	13.49	111.1	17.64	139.7	22.18
116	3.082	−18.0	−2.86	19.4	3.06	40.4	6.42	63.2	10.03	88.0	13.97	115.1	18.28	144.7	22.96
117	3.170	−18.7	−2.97	20.1	3.19	41.8	6.64	65.4	10.38	91.2	14.49	119.2	18.94	150.0	23.83
118	3.260	−19.3	−3.07	20.8	3.31	43.3	6.88	67.8	10.77	94.4	15.00	123.5	19.63	155.4	24.70
119	3.353	−20.0	−3.18	21.6	3.43	44.9	7.14	70.3	11.18	97.9	15.56	128.0	20.35	161.1	25.61
120	3.448	−20.7	−3.29	22.4	3.56	46.6	7.41	72.8	11.58	101.4	16.13	132.7	21.10	167.1	26.58
121	3.545	−21.4	−3.40	23.2	3.77	48.3	7.68	75.5	12.01	105.1	16.72	137.6	21.89	173.2	27.56
122	3.644	−22.2	−3.53	24.0	3.82	50.0	7.96	78.2	12.45	109.0	17.35	142.6	22.70	179.6	28.58
123	3.746	−23.0	−3.66	24.9	3.96	51.8	8.25	81.1	12.91	112.9	17.98	147.9	23.55	186.3	29.66
124	3.850	−23.8	−3.79	25.8	4.11	53.7	8.55	84.0	13.38	117.1	18.65	153.3	24.42	193.2	30.77
125	3.956	−24.7	−3.94	26.7	4.25	55.6	8.86	87.1	13.88	121.4	19.34	159.1	25.35	200.5	31.95
126	4.065	−25.6	−4.08	27.6	4.40	57.7	9.20	90.3	14.39	125.9	20.07	165.0	26.30	208.0	33.15
127	4.177	−26.5	−4.23	28.6	4.56	59.8	9.54	93.6	14.93	130.6	20.83	171.2	27.30	215.8	34.41
128	4.291	−27.5	−4.39	29.7	4.74	62.0	9.89	97.1	15.49	135.5	21.61	177.6	28.33	224.0	35.73
129	4.408	−28.5	−4.55	30.8	4.92	64.3	10.26	100.7	16.07	140.6	22.44	184.3	29.41	232.6	37.12
130	4.527	−29.5	−4.71	32.0	5.11	66.7	10.64	104.5	16.68	145.9	23.29	191.4	30.55	241.5	38.55
131	4.650	−30.6	−4.89	33.2	5.30	69.2	11.05	108.5	17.33	151.4	24.18	198.7	31.73	250.8	40.05
132	4.775	−31.8	−5.08	34.4	5.50	71.8	11.47	112.6	17.99	157.2	25.11	206.3	32.96	260.6	41.63
133	4.903	−32.9	−5.26	35.8	5.72	74.5	11.91	116.9	18.68	163.3	26.10	214.3	34.25	270.8	43.28
134	5.034	−34.2	−5.47	37.1	5.93	77.4	12.37	121.4	19.41	169.6	27.12	222.7	35.60	281.4	44.99
135	5.168	−35.5	−5.68	38.5	6.16	80.3	12.84	126.0	20.15	176.2	28.18	231.4	37.01	292.6	46.80
136	5.305	−36.8	−5.89	40.0	6.40	83.4	13.34	130.9	20.94	183.1	29.30	240.5	38.48	304.2	48.67
137	5.445	−38.2	−6.11	41.5	6.64	86.6	13.86	136.0	21.77	190.3	30.46	250.1	40.03	316.4	50.64
138	5.588	−39.7	−6.36	43.2	6.92	90.0	14.41	141.4	22.64	197.8	31.67	260.1	41.65	329.3	52.73
139	5.735	−41.2	−6.60	44.8	7.18	93.6	14.99	147.0	23.55	205.7	32.95	270.6	43.34	342.7	54.89
140	5.884	−42.8	−6.86	46.5	7.45	97.3	15.59	152.8	24.48	214.0	34.29	281.6	45.12	356.8	57.17

t = Dry bulb temperature (F).
t' = Wet bulb temperature (F).
p = Barometric pressure (in. of Hg).
Δp = Pressure difference from standard barometer (in. of Hg).
W = Moisture content of air (gr per lb of dry air).
W_s' = Moisture content of air saturated at wet bulb temperature t' (gr per lb of dry air).
ΔW = Moisture content correction of air when barometric pressure differs from standard barometer (gr per lb of dry air).
$\Delta W_s'$ = Moisture content correction of air saturated at wet bulb temperature when barometric pressure differs from standard barometer (gr per lb of dry air).

NOTE: To obtain ΔW reduce value of $\Delta W_s'$ by 1% where

$t - t' = 24$ F and correct proportionally when $t - t'$ is not 24 F.

h = Enthalpy of moist air (Btu per lb of dry air).
Δh = Enthalpy correction when barometer pressure differs from standard barometer, for saturated or unsaturated air. (Btu per lb of dry air).
v = Volume of moist air (cu ft per lb of dry air).

$$= \frac{.754\,(t + 459.7)}{p}\left[1 + \frac{W}{4360}\right]$$

Example: At a barometric pressure of 25.92 with 220 F DB and 100 F WB, determine W, h, and v. $\Delta p = -4$ and from table $\Delta W_s' = 50.4$. From note above,

$$\Delta W = \Delta W_s' - \left(\frac{120}{24} \times .01 \times 50.4\right) = 50.4 - 2.5 = 47.9$$

Therefore W = 102 (from chart) + 47.9 = 149.9 gr per lb of dry air. From table Δh = 7.95. Therefore h = saturation enthalpy from chart + deviation + 7.95 = 71.7 − 2.0 + 7.95 = 77.65 Btu per lb of dry air. From equation above

$$v = \frac{.754\,(220 + 459.7)}{25.92}\left[1 + \frac{149.9}{4360}\right] = 20.43 \text{ cu ft per lb of dry air}$$

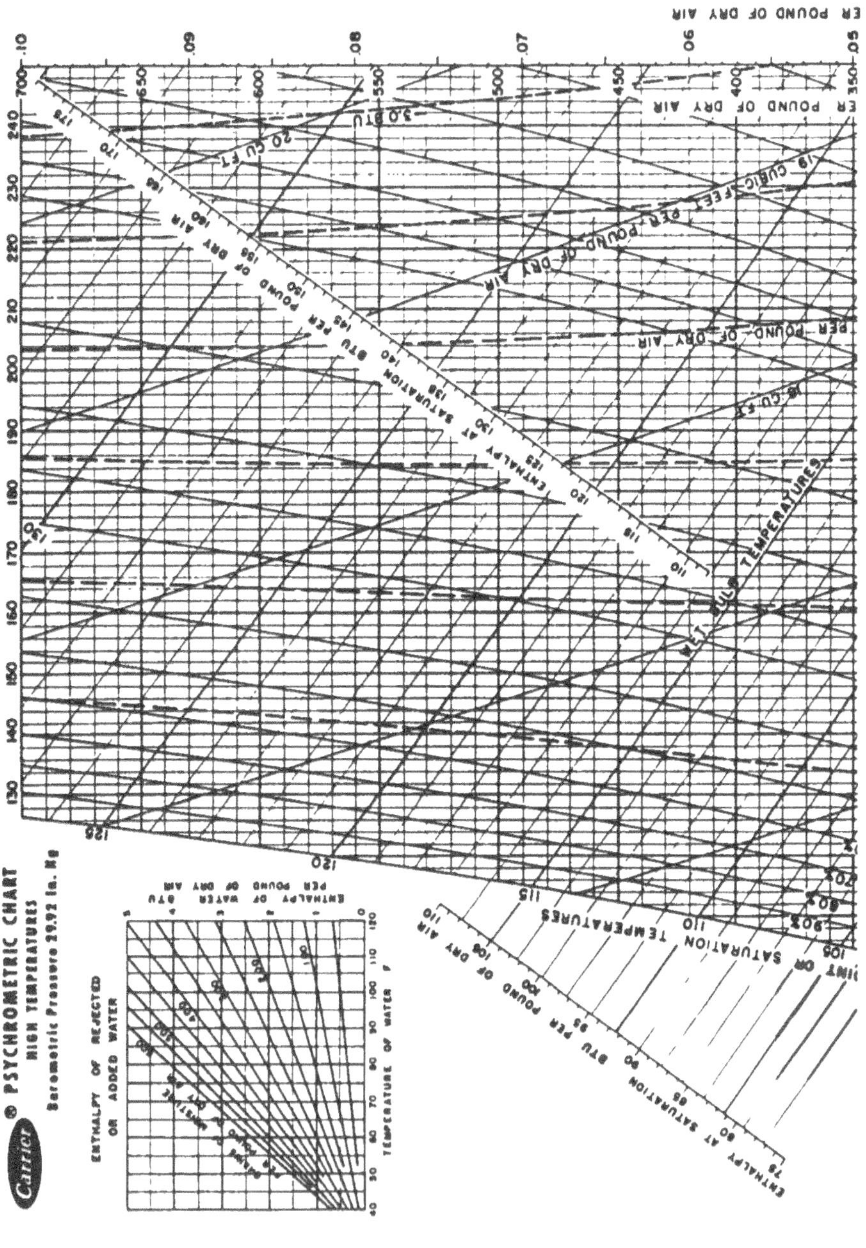
Carrier
PSYCHROMETRIC CHART
HIGH TEMPERATURES
Barometric Pressure 29.92 In. Hg
ENTHALPY OF REJECTED OR ADDED WATER
ENTHALPY OF WATER BTU PER POUND OF DRY AIR
TEMPERATURE OF WATER F
ENTHALPY AT SATURATION BTU PER POUND OF DRY AIR
DEW POINT OR SATURATION TEMPERATURES
WET BULB TEMPERATURES
18 CU FT
PER POUND OF DRY AIR
19 CUBIC FEET PER POUND OF DRY AIR
20 CU FT
3.0 BTU
PER POUND OF DRY AIR

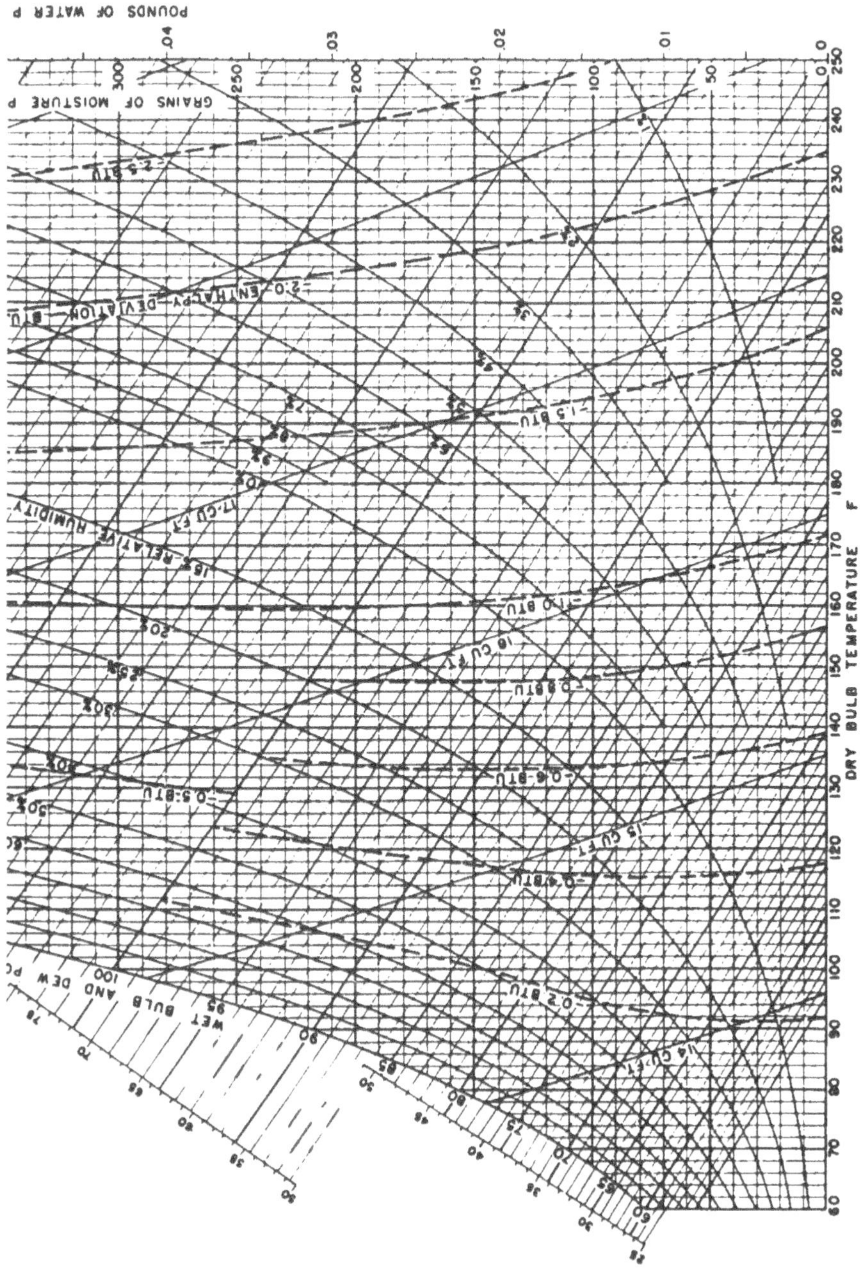
DRY BULB TEMPERATURE F
WET BULB AND DEW PO
GRAINS OF MOISTURE P
POUNDS OF WATER P
15% RELATIVE HUMIDITY
2.0 ENTHALPY DEVIATION BTU
14 CU FT
15 CU FT
16 CU FT
17 CU FT

REFERENCES

1. Test Methods for Evaluating Solid Waste, SW-846, 3rd ed., USEPA, Office of Solid Waste and Emergency Response, Washington, D.C., 11/86.
2. Code of Federal Regulations, Title 40, 40CFR60, 40CFR50.
3. Burghardt, M.D., Engineering Thermodynamics with Applications, 3rd ed., Harper & Row, New York, 1986.
4. Obert, E.F., *Internal Combustion Engines and Air Pollution*, Intext Educational Publishers, New York, 1973.
5. Cooper, C.D., and Alley, F.C., *Air Pollution Control: A Design Approach*, Waveland Press, Prospect Heights, IL, 1986.
6. Brenchley, D.L., Turley, C.D., and Yarmac, R.F., *Industrial Source Sampling*, Ann Arbor Science, Ann Arbor, MI, 1973.
7. Stevens, R.K., and Herget, W.F., *Analytical Methods Applied to Air Pollution Measurements*, Ann Arbor Science, Ann Arbor, MI, 1974.
8. Nelson, G.O., *Controlled Test Atmospheres*, Ann Arbor Science, Ann Arbor, MI, 1971.
9. Nagda, N.L., Rector, H.E., and Koontz, M.D., *Guidelines for Monitoring Indoor Air Quality*, Hemisphere, Washington, D.C., 1987.
10. Keith, L.H., *Principals of Environmental Sampling*, American Chemical Society, Washington, D.C., 1987.
11. Winberry, W.T., Murphy, N.T., and Riggan, R.M., *Methods for Determination of Toxic Organic Compounds in Air*, Noyes Publishing, Park Ridge, NJ, 1990.
12. Quality Assurance Handbook for Air Pollution Measurement Systems, USEPA EPA/600/7-77/027b, Vol. III, Stationary Source Specific Methods Environmental Monitoring Systems Laboratory, Research Triangle Park, NC, 1986.

13. Rom, J.J., Maintenance, Calibration, and Operation of Isokinetic Source Sampling Equipment, USEPA APTD-0576, Office of Air Programs, Applied Technology Division, Research Triangle Park, NC, 1972.
14. APTI Course 450, Source Sampling for Particulate Pollutants, Student Manual, USEPA EPA/450/2-79/006, Air Pollution Training Institute, Research Triangle Park, NC, 1979.
15. APTI Course 435, Atmospheric Sampling, Student Manual, USEPA EPA/450/2-80/004, Air Pollution Training Institute, Research Triangle Park, NC, 1980.
16. Ruby, M.G., Katchamat, S., and Wangwongwatana, S., Air Pollution Measurements Laboratory Lab Manual, University of Cincinnati, Cincinnati, OH,1982
17. APTI Course SI:483A, Measuring the Emission of Organic Compounds to the Atmosphere, USEPA EPA/450/2-84/005, Air Pollution Training Institute, Research Triangle Park, NC, 1979.
18. Smith, W.S., Martin, R. M., Durst, D.E., Hyland, R.G., Logan, T.J., and Hagar, C.B., "Stack Gas Sampling Improved and Simplified with New Equipment", Paper 67-119, 58th Annu. Meet. of Air Pollution Control Assoc., June, 1967.
19. Godish T., *Air Quality*, 2nd ed., Lewis, Chelsea, MI, 1991.
20. Hesketh, H., *Air Pollution Control, Traditional and Hazardous Pollutants*, Technomic Publishing, Lancaster, PA, 1991.
21. Ness, S.A., *Air Monitoring for Toxic Exposures, An Integrated Approach*, Van Nostrand Reinhold, New York, 1991.
22. Flagan, R.C., and Seinfeld, J.H., *Fundamentals of Air Pollution Engineering*, Prentice Hall, Englewood Cliffs, NJ, 1988.
23. Mitchell, W.J., and Midgett, M.R., "Field Reliability of the Orsat Analyzer", *JAPCA*, 26(5), 491, 1976.
24. Nolan, M., and Marshalla, A., *Pollut. Eng.*, 5:35, 1973.
25. Martin, R.M., Construction Details of Isokinetic Source-Sampling Equipment, USEPA APTD 0581, Research Triangle Park, NC, 1971.
26. Jutze, G.A., and Foster, K.E., "Recommended Standard Method for Atmospheric Sampling of Fine Particulate Matter by Filter Media — High-Volume Sampler", *JAPCA*, 17(1), 17, 1967.
27. Lodge, J.P., Jr., Ed., *Methods of Air Sampling and Analysis*, Lewis, Chelsea, MI, 1989.
28. Measurement of Toxic and Related Air Pollutants, Proc. EPA/AWMA Int. Symp., 1986–1992, RTP, NC, AWMA, Pittsburgh, PA.
29. Silverman, L., and Viles, F.G., "A High Volume Air Sampling and Filter Weighing Method for Certain Aerosols", *J. Ind. Hygiene Toxicol.*, 30, 124, 1948.
30. Gerstle, R.W., "Finding a Solution to Your Air Pollution", *Pollut. Eng.*, 24, 11, 1992.
31. Edwards, L.E., and Nottoli, J.A., "Source Sampling Tests Stack Emissions", *Environmental Protection*, 3, 5, June 1992.

32. Baker, W.C., and Pouchot, J.F., "The Measurement of Gas Flow, Part I", *JAPCA*, 33(1), Jan 1983.
33. Baker, W.C., and Pouchot, J.F., "The Measurement of Gas Flow, Part II", *JAPCA*, 33(2), Feb 1983.
34. NIOSH Manual of Analytical Methods, 3rd ed., (84–100) plus supplements, National Institute for Occupational Safety and Health, Cincinnati, OH, 1984.
35. *OSHA Analytical Methods Manual*, with updates, GPO: Washington, DC, 1985.
36. Betz, W.R., Wachob, G.D., and Firth, M.C., "Monitoring a Wide Range of Airborne Organic Contaminants", in Measurement of Toxic and Related Air Pollutants, Proc. 1987 EPA/AWMA Symp., APCA, Pittsburgh, PA, 1987.
37. Riggan, R.M., Technical Assistance Document for Sampling and Analysis of Toxic Organic Compounds in Ambient Air, EPA-600/4-83-027, U.S.EPA, Research Triangle Park, NC, 1983.
38. Harrison, R.M., and Perry, R., *Handbook of Air Pollution Analysis*, 2nd ed., Chapman and Hall, London, 1986.
39. Loo, B.W., Jaklevic, J.M., and Goulding, F.S.; Dichotomus Virtual Impactors for Large Scale Monitoring of Airborne Particulate Matter, presented at the Symp. on Fine Particles, May 1975.
40. Air Sampling Instruments for Evaluation of Atmospheric Contaminants, Am. Conf. of Governmental Industrial Hygienists, Cincinnati, OH, 1978.
41. Cadle, R.D., *The Measurement of Airborne Particles*, Wiley-Interscience, New York, 1975.
42. Quality Assurance Handbook for Air Pollution Measurement Systems, USEPA EPA/600/7-77/027a, Vol. II, Ambient Air Pecific Methods Environmental Monitoring Systems Laboratory, Research Triangle Park, NC, 1986.
43. Wedding, J.B., and Weigand, M.A., Operations and Maintenance Manual, The Wedding Associates PM10 Critical Flow High-Volume Sampler, Fort Collins, CO, 1987.
44. Lodge, J.P., Pate, J.B., Ammons, B.E., and Swanson, G. A., "The Use of Hypodermic Needles as Critical Orifices in Air Sampling", *JAPCA*, 16(4), April 1966.
45. Corn, M., and Bell, W., "A Technique for Construction of Predictable Low Capacity Critical Orifices", *Am. Ind. Hygiene Assoc. J.*, 24, 502, 1963.
46. Perry, J.H., *Chemical Engineers' Handbook*, McGraw-Hill, New York, 1950.
47. Marks, L.S., *Mechanical Engineers Handbook*, 35th ed., McGraw Hill, New York, 1982.
48. Jahnke, J.A., *Continuous Emission Monitoring*, Van Nostrand Reinhold, New York, 1993.

INDEX

D

E

F